Roofing and Insulation

ROOFING
& INSULATION

Mike Lawrence

The Crowood Press

First published in 1991 by
The Crowood Press Ltd
Ramsbury, Marlborough,
Wiltshire SN8 2HR

www.crowood.com

New edition 2001

This impression 2006

British Library Cataloguing-in-Publication Data
A catalogue record for this book is available from the British Library.

ISBN 1 86126 464 X
EAN 978 1 86126 464 0

Acknowledgements

Line-drawings by Andrew Green

The author would like to thank the following organizations for providing the photographs listed below:

Blue Hawk (pages 11 and 91);
Evo-Stik (pages 31, 35 and 92);
Marshall Cavendish (pages 8, 12, 19, 25, 37, 41–53, 63, 71, 73–5, 85, 88 and 89);
Redland Roof Tiles (page 10);
Robert Harding Picture Library (pages 13–17, 27, 69 and 87);
Rockwool (pages 55–61, 65 and 73);
3M UK (pages 78–81).

Typeset by Acūté, Stroud, Glos
Printed and bound by Times Offset (M) Sdn Bhd, Malaysia

Contents

Introduction 6

THE BASICS

Surveying your house 7
Raw materials 10
Shopping and safety 18
Access equipment 19
Using ladders safely 20
Using access towers 22
Working on the roof 24

ROOFING JOBS

Replacing tiles 26
Replacing slates 28
Patching a flat roof 29
Resurfacing a flat roof 30
Repairing a glass roof 32
Repairing a corrugated roof 33
Repairing flashings 34
Repairing chimney-stacks 36
Dealing with disused flues 38
Repairing eaves woodwork 40
Repairing gutters 42
Repairing downpipes and hoppers 44
Repairing valley gutters 46

WEATHERPROOFING JOBS

Replacing putty 48
Repairing leaded lights 50
Sealing round frames 51
Repairing defective brickwork 52
Clearing DPC bridges 54

INSULATING JOBS

Laying blanket loft insulation 56
Laying loose-fill loft insulation 58
Improving loft ventilation 59
Insulating roof slopes and loft rooms 60
Insulating flat roofs 62
Insulating exterior walls 64
Insulating tanks and cylinders 67
Insulating pipework 70
Insulating floors 72
Draught-proofing doors and windows 74
Double glazing 79
Sound-proofing 82
Fitting ventilators 84

EXPENSIVE JOBS

Finding a contractor 85
Quotations 86
Getting professional help 87

FACTS AND FIGURES

Tiles and slates 89
Flat roof materials 90
Damp-proofing materials 91
Insulation materials 92
Draught-proofing materials 93

Glossary and Useful Addresses 94
Index 95

Introduction

As home ownership rises, more and more people now have a serious interest in keeping their homes in good order. After all, it is likely to be the biggest investment of their lives, and no one wants to preside over a deteriorating asset. Unfortunately, our climate has other ideas, and our homes are under constant everyday attack from a combination of heat, cold, wind and rain, all trying to get in while we do our best to keep them out.

Keeping a house weatherproof and in good condition is an ongoing task; there is always something to attend to, and it is better to carry out regular maintenance on a rolling basis than to let things go to the point where major – and frequently expensive – repairs become necessary. To do this you need to be something of a jack of all trades – tiler, slater, steeplejack, bricklayer, plumber, glazier – even if you are master of none. This book will be your guide. It tackles a wide range of roofing and weatherproofing jobs, any of which you may have to carry out at some time on your house. It tells you about the materials you will have to buy and the techniques you will need to employ – all well within the capabilities of the average do-it-yourselfer. It also stresses the importance of safety when working outdoors.

It also deals with the subject of insulation, which is the art of keeping warm when you want to be, and every home's vital contribution to cutting wastage of valuable fuel resources. Put them all together, and you will have a warm, dry, sound house to live in.

THE BASICS

The typical British home is a complicated structure, made up from a variety of materials. Its roof may be sloping and covered with tiles, slates, stones or man-made materials, or flat and finished with felt, asphalt or metal sheeting. Its walls may be of brick or stone, left bare or covered with rendering, tiles or timber cladding to make them more weatherproof. Set into them are door and window openings of all sorts of shapes and sizes, filled with frames in stone, timber or metal. All try to coexist in keeping the building's occupants warm and dry, but the very complexity of the structure make this a tall order. The elements gradually wear away the structure and penetrate any weak spots, letting wind and water in and heat out, and eventually leading to serious decay. Keeping the elements at bay is really what exterior home maintenance is all about.

The only way to understand what you are up against is to make yourself familiar with every individual element of your home's construction, from the top of the chimney-stacks to where the house walls meet the ground, by carrying out a detailed inspection. The exercise is rather like building up a service history for your car, in order that you know how everything works and what needs repairing or replacing to keep it running sweetly.

Start at the Top

You may be able to inspect your roof and chimney-stacks from a vantage point such as a neighbour's upstairs window, but there is really no substitute for climbing a ladder at least to eaves-level so you can make your inspection at close quarters from every side of the roof. Read the advice on pages 20–21 before you climb, so you know how to set up and use ladders safely. Falls account for thousands of injuries and dozens of deaths in the home every year, and you do not want to be among them.

Fig 1 (*below*) The exterior of a typical home features a wide variety of different materials, and only if all are maintained in good condition can the overall structure remain weatherproof.

When you climb your ladder, take a pair of binoculars with you if you own or can borrow them, as they make it easier to inspect things like the mortar pointing on chimney-stacks or the condition of the metal strips (called flashings) that seal the joints between chimney-stacks and the roof slope. Carry a notebook and pen too. Once you are at eaves-level, run through the following questions, and make notes of the answers. They will form the basis for your future maintenance programme, as well as telling you things you may not have known about your house.

The chimney-stack Is the stack itself perfectly vertical, or are there signs of leaning in any direction? Is the pot sound or cracked? Has it been removed if the stack is no longer used? If so, has the flue been capped in any way? Is the mortar (called flaunching) that secures the pot to the stack still in good condition, or is it cracked and lifting away from the masonry? Is the stack brickwork in good condition, or are there signs of frost damage to the brick faces and corners? Is the pointing intact? If the stack is rendered, is this sound, or has it cracked and broken away? If a TV aerial is mounted on the stack, are its fixings secure, or is the mast leaning because they have failed?

The stack/roof junction If metal flashings have been used to waterproof this point in the structure, are they intact? Are the upstanding parts properly tucked into the mortar courses of the stack brickwork, or have they pulled out? Are the parts that rest on the roof slope flat on the tiles or slates, or are there signs of edges lifting and being torn by high winds? If mortar flashings have been used instead of metal, are they sound, or is there evidence of cracking and failure?

The roof ridge Are the tiles along the ridge of the roof properly bedded on mortar, with intact pointing between them, or do they appear to have dry joints? If the roof has a hip (the external angle where two adjacent roof slopes meet), are the ridge tiles secure here? Are any tiles cracked or missing altogether?

The roof slopes Is everything intact, or are there signs of tiles or slates that are cracked or have slipped out of position? Are any missing altogether? At gable ends above triangular sections of the house wall, are the ends of the tile courses soundly bedded on mortar? If there are valleys (internal angles between adjacent roof slopes), are these formed with tiles or lined with metal sheeting? If the former, are the tiles intact, or have any slipped out of place? If the latter, is the valley clear of debris? Are the tiles overlapping it securely bedded on mortar pointing, or are there gaps?

Fig 2 (*above*) A modern pantiled roof with overhanging eaves and mortared verges. Note the lead flashings where the lower roof abuts the house wall.

The eaves Are the gutters level, clear of debris and in good condition, or are there blockages, sagging areas and signs of rust attack? Are plastic gutters intact, or are there splits and cracks in places? Is the eaves woodwork sound and well decorated, or is paint peeling to reveal patches where rot has taken hold? If there are barge-boards at gable ends, are these sound or decaying?

Flat roofs If your house has any flat roofs, it should be simple enough to climb on to them so you can check their condition. If the roof is felted, is its covering layer of protective chippings still in place, or have they all migrated elsewhere long ago? Are there signs of the felt lifting at seams or edges? Are there any cracks or blisters? On metal roofs, is the decking flat and intact, or have seams and joints lifted or opened up? If the roof meets a house wall, are the flashings along the junction intact and bonded to the wall, or are their tears and signs of lifting?

Lean-to roofs If you have shallow sloping roofs over porches, bay windows and the like, is the roof covering in good condition? Are flashings intact where the roof slope meets the house wall?

Work down the Walls

Once you have completed your high-level inspection, shorten the ladder and give each outside wall a once-over in the same way. Again, here are the questions you need to answer.

Masonry walls Is fair-faced brick or stone in good condition, or has frost damage broken away the faces? Is pointing sound, or are there cracks and sections missing that could allow water penetration? Are there signs of staining or mould growth on the wall surface? Are there any cracks through bricks, rather than along mortar courses?

Rendered walls Is rendering and pebbledashing sound, or is the surface cracked and chalky to the touch? Are there patches that sound hollow when tapped, or are missing altogether?

Downpipes and soil pipes Are these in good condition and securely fixed to the wall, or are there signs of leaks, rust and physical damage?

Windows and doors Are frames sound and well decorated, or is paintwork peeling to reveal patches of rotten wood or rusty metal? Is the junction between frame and masonry well sealed, or are there gaps that could let water or draughts in? Is the putty around individual frames sound and intact, or are there patches that are cracked, lifting away from the glass or missing altogether? Do leaded lights leak?

Ground level Are flower-beds, paths, steps and patio surfaces less than 150mm (6in) below the level of the damp-proof course (DPC) in the house wall, allowing rainwater to splash up the walls and cause damp to penetrate? Similarly, are there any garden walls or outbuildings constructed against the house wall which could bridge the DPC?

Once you have completed your inspection of the house exterior, you should have a detailed picture not only of how it is constructed, but also of what is needed to put right any faults that you have discovered. The other half of the overall weatherproofing equation concerns your home's ability to retain heat – in other words, how good its insulation and draught-proofing are performing. Here again, the best solution is to check over what you have and record what needs to be done.

An Insulation Check-List

As with your exterior inspection, start at the top of the house.

The loft Does the loft have adequate insulation on its floor, or on the underside of the roof slope if there are rooms in the roof space? Are any tanks or pipes in the loft well insulated? Is there adequate ventilation of the loft space to prevent condensation? Is the loft hatch insulated and draught-proofed?

The external walls Is there cavity wall insulation? Is there any other insulation of exterior walls? If not, are exterior walls of cavity or solid construction?

Windows and doors Do windows and exterior doors close properly in their frames? Are they all draught-proofed? Do windows suffer from condensation? Is there any double glazing? If so, is it effective, or does it suffer from condensation problems too?

Floors Do draughts rise through suspended timber floors? Are solid floors cold and prone to condensation? Are underfloor ventilators clear and working?

Fig 3(*below*) Exterior walls offer plenty of opportunities for weather damage and water penetration – frosted bricks, missing pointing, gaps around door and window frames, even damp caused by badly positioned walls and patio surfaces.

The raw materials you need to keep your house weatherproof and warm fall into five broad groups: tiles and slates for pitched roofs, coverings for flat roofs, liquid damp-proofers and related repair materials, insulation, and draught-proofing.

Roof Tiles

Tiles are the most popular covering for sloping (pitched) roofs, and with good reason. As well as being durable, easy to fix in place and good at keeping out the weather, they come in a wide range of colours and designs.

Clay tiles are the traditional choice, and are undeniably attractive despite the fairly limited range of plain and pantile shapes available. There is generally a choice of plain, sand-faced and brindled (streaky) finishes. In addition to the basic tile, there are matching valley and bonnet hip tiles, allowing the tile coursing to be maintained through the interior and exterior angles between one roof slope and an adjacent one without the need to use ridge-tiles or metal valley gutters.

For most roofs on new houses, concrete tiles are widely used. They come in plain and interlocking profiles in a wide range of colours and finishes, mainly greys, reds, oranges and browns. Their only visual drawback is that they are too uniform in colour for some tastes; clay tiles all vary slightly in colour and so provide a more natural look to the roof surface.

Concrete tiles are also generally heavier then many old roof coverings, so fitting them to an existing roof in place of a lighter tile may entail strengthening of the roof structure. However, interlocking types offer excellent weatherproofing as well as fast installation, and all types promise great durability.

Most tile ranges have complementary ridge-tiles available, including types with built-in ventilators to provide essential ventilation to the roof timbers and the loft space. Some specialist firms also offer ornamental ridge-tiles that are ideal for repair and restoration work on Victorian and Edwardian houses.

Fig 4 (*below*) Interlocking roof tiles are widely used on new homes, while plain clay tiles have their place for repair work. *See* page 89 for more details.

Slates

Natural slate is the traditional pitched roof covering that has all but priced itself out of the market. New slates are still sold, but you may have trouble finding a local stockist and may have to join a long queue of waiting buyers. Second-hand slates are a better bet for repair work, so long as you can find them in a thickness to match those already on your roof, but they need checking carefully for defects.

An alternative, especially for replacement work, is the man-made slate. Older asbestos-cement types have been superseded by fibre-cement and interlocking concrete slates, and by extremely realistic-looking reconstituted slates made by binding crushed slate together with synthetic resins. However, like concrete tiles, they tend to lack the subtle colour variations of natural slate.

Apart from the plain rectangle, man-made slates are also available with rounded, pointed or chisel-shaped ends, which can create a highly dramatic effect.

Ridges are finished off with special-shaped ridge-tiles, coloured to match the slates. They can be used on hips too, although here a mitred finish with metal soakers underneath as waterproofing gives a neater finish.

Roofing Felts

Whether you are dealing with a flat-roofed extension to your home or just the garden shed, some sort of roofing membrane (what most people call roofing felt) provides the simplest way to keep out the rain.

Roofing felt consists of a fabric core which is saturated and coated with bitumen to make it waterproof. Plain types are used as underlayers in built-up flat-roof systems, which are used on extensions, for example. They also serve as the top layer (or capsheet) if this is then protected with mineral chippings laid in more bitumen. Types with a surface coating of fine mineral granules are used as an alternative capsheet of a built-up roof if chippings are not laid, and also on their own as a single-layer pitched- or flat-roof covering for outbuildings.

Roofing felt is classified under British Standard 747 into five groups. Class 1 felts have a vegetable-fibre base, and are the cheapest and best used only on outbuildings, although they can be used in built-up flat roofs as well, if low cost is more important than long life. They are also used as underfelting on pitched roofs.

Class 2 felts are similar but incorporate asbestos fibres as reinforcement and fireproofing, and are also used as a single layer or in built-up roofs.

Class 3 felts have a glass-fibre base that is completely rot-proof and which can cope better than other types with movement in the roof surface, since it resists tearing and splitting well. This type is, needless to say, more expensive that Class 1 or 2 felts.

Class 4 felts are not used as waterproof roof coverings; instead they form a sheathing layer under asphalt and metal roof coverings and are the province of the professional roofing contractor.

Class 5 felts have a polyester-fibre base which is the strongest of all BS felts and also the most expensive at around twice the price of Class 3 felt. There is also a range of high-performance roofing felts with a glass-fibre and polyester base, not included in the British Standard classification.

Fig 5 (*below*) Roofing felts come in a range of grades and weights, and it is vital to choose the right one for a particular roofing job. *See page 90 for more details.*

Damp-Proofers and other Repair Materials

You will need a range of materials for jobs as diverse as waterproofing flat roofs, patching flashings, filling cracks and sealing porous masonry.

Liquid damp-proofers are one of the most useful groups of products, and fall into two broad categories. The first are the bituminous sealers, used for repairs to flat roofs, flashings and valley gutters, while the second are the silicone-based water repellents employed mainly for curing porous brickwork and rendering. Both are designed to be brushed onto the surfaces being treated.

To cope with defects requiring something more substantial than a liquid sealer, you can turn to the various types of repair tape or reinforcing membrane. Self-adhesive flashing tape is useful for a wide range of roof repairs, and can be pressed into service for fixing leaky gutters and downpipes as well. Clear repair tape can cope with cracked glass or plastic sheet.

Fig 6 (*above*) You will need liquid damp-proofer for several outside jobs, plus repair tapes and reinforcement for bridging gaps and major defects in a variety of surfaces.

Fig 7 (*left*) Non-setting mastic is useful for jobs like sealing gaps between door and window frames and the surrounding masonry, while putty is essential for keeping windows weatherproof.

Raw Materials

Mastics of various types are used for filling and waterproofing gaps and making good damage. Use bituminous types for patching up cracks in flat-roof surfaces and for repairing leaky joints in cast-iron guttering and downpipes, and exterior-quality non-setting mastics – the type sold in cartridges and dispensed from a cartridge gun – for jobs like sealing gaps around window frames. You will also need putty of the appropriate formulation to weatherproof timber and metal windows.

Tools

You will need few specialist tools for the weatherproofing jobs described here – just an assortment of general-purpose tools plus such basic masonry tools as a cold chisel, club hammer and trowel. The one tool you may need to hire is a slater's ripper, used for cutting through the nails holding tiles and slates in place on pitched roofs. It is also a good idea to invest in a good-quality tool belt or apron; you will be working off a ladder for many jobs, and this gives you somewhere safe to store tools and materials.

Insulation Materials

When it comes to keeping the heat in instead of the weather out, you will need a range of different insulating materials to cope with the requirements of different areas of the house structure.

Loft insulation This is one of the most important, because as much as 25 per cent of the total heat lost from a poorly insulated house leaves through the roof space, and also because this is an area that is easy to insulate relatively inexpensively.

The two most widely used loft insulation materials are glass fibre and mineral wool, both of which are completely inert materials that have excellent insulation properties. Both are available in the form of blankets which are designed to be unrolled between the loft floor joists. Mineral wool is also sold as loose-fill material, which is useful in lofts full of awkward obstructions, and in the form of semi-rigid insulation

Fig 8 (*left*) Blanket-type glass-fibre or mineral-wool materials are the most widely used for loft insulation, but loose-fill materials such as expanded vermiculite are useful in lofts that are full of obstructions. *See* page 92 for more details of available sizes and coverages.

batts, which can be used for wall, ceiling and under-floor insulation too. They should be laid to a minimum depth of 100mm (4in), and to 150mm (6in) if the joist depth allows.

An alternative to fibre insulation is to use granules of an expanded mineral called vermiculite. These are poured out between the joists up to the level of their top edges, but they do not provide as much insulation, inch for inch, as glass fibre or mineral wool and need covering and restraining at the eaves so that they do not blow about in draughty lofts.

All these materials can also be used for insulating flat roofs, although they are difficult to add once the roof has been constructed. So long as access can be gained to the void between roof deck and ceiling, it is generally easier to insert slabs of rigid polystyrene insulation between the roof joists.

If you prefer not to lay your own loft insulation, you can have mineral wool or specially treated cellulose fibres blown into your loft space by professional installers, but this will obviously be more expensive that a DIY installation.

Wall insulation This is the next priority. Exterior walls are also responsible for a high percentage of heat losses – up to 35 per cent of the total in some cases – so insulation is obviously of benefit here too. Homes with cavity walls can have the cavities filled with a variety of insulation materials including urea-formaldehyde foam and mineral wool, but this is a job for a professional installer and also needs Building Regulations approval from your local-authority building inspector. However, if you are building an extension from scratch, you can insulate the new walls by incorporating semi-rigid insulation batts in the cavity as the walls are built up.

Homes with solid exterior walls can be

Fig 9 (*left*) Cavity walls can be insulated by filling the cavities with foam or chopped fibre insulation (top), but this is a job for professional installers. Alternatively the inner face of cold exterior walls can be lined with insulating plasterboard (centre), or with ordinary plasterboard over a layer of semi-rigid insulation batts or rigid polystyrene.

treated, again professionally, by fitting exterior cladding, but this is expensive and noticeably alters the exterior of the house. A better bet is to line interior walls with insulation, either blanket or semi-rigid batts fitted between battens and covered with plasterboard, or else special insulating plasterboard, which can be fixed direct to the wall surface.

Tank and pipe insulation These are important more in the context of preventing freeze-ups in cold weather than in controlling heat loss, although insulating hot-water cylinders and central-heating pipework will make a worthwhile contribution to avoiding unnecessary waste and are inexpensive to buy and install.

For storage and feed-and-expansion tanks in loft spaces, you can either use loft insulation materials to make up an insulating cover or buy a proprietary kit, usually of rigid polystyrene for rectangular tanks, or plastic-wrapped glass fibre or mineral wool for round ones. Proprietary jackets are also available for all common sizes of cylindrical hot-water cylinder.

As far as pipe insulation is concerned, you have a choice of using pipe bandage, which is wound around the pipe and tied in place, or various types of split sleeve insulation, which are slipped into place along the length of the pipe runs. Bandages are made of mineral wool or glass fibre, and may be encased in, or faced with, plastic, while foam types may be flexible or semi-rigid and come in different sizes to suit the most commonly used pipe diameters on domestic plumbing and heating systems – 15, 22 and 28mm.

Fig 10 (*above*) Proprietary jackets are available in a range of sizes to cut heat losses from hot cylinders, and also to prevent freeze-ups in storage tanks and feed-and-expansion tanks in insulated lofts.

Fig 11 (*left*) Pipe bandage or slip-on pipe sleeves are essential for insulating both hot and cold pipes – the former to reduce heat losses and the latter to prevent freeze-ups.

Draught-Proofing

Draught-proofing is a form of indirect insulation that does its job by preventing cold draughts entering the house and forcing expensively heated air out of it. Since up to 15 per cent of all heat losses can occur as a result of ill-fitting doors and windows, and draught-proofing materials are relatively inexpensive, it makes sound sense to fit draught excluders to all your doors and windows, and also to tackle other sources of draughts like letter-box openings and loft hatch-doors.

Doors and opening windows must have some clearance to allow them to open and shut without binding in their frames. The various types of draught excluder on the market aim to block that gap by providing a flexible seal.

For **door bottoms**, there are three main types of excluder. The first fits along the bottom face of the door, on either the inside or outside face, and carries a brush or flap seal that closes against the sill as the door closes. The second fits across the door threshold itself, and is compressed as the door edge closes onto it. The third is a two-part excluder; one section is attached to the bottom of the door's outer face, while the other fits across the threshold, and the two interlock when the door is closed to provide a draught-proof and weatherproof seal.

For **door frames and casement windows**, there is a wider range of excluders to choose from. The simplest are self-adhesive rubber or foam seals which are stuck round the rebate and are compressed as the door or window closes onto them. More durable, but fiddlier to fit, is the V-seal type, which is a strip of nylon or sprung metal that is pinned or stuck to the rebate around the frame and presses against the edge of the door or window rather than its face. Most expensive is the brush pile or synthetic rubber tube seal held in a rigid moulding, designed to be fixed to the outside of the frame of inward-opening doors and to the inside for outward-opening windows. It is also invaluable for draught-proofing the sides of sliding windows.

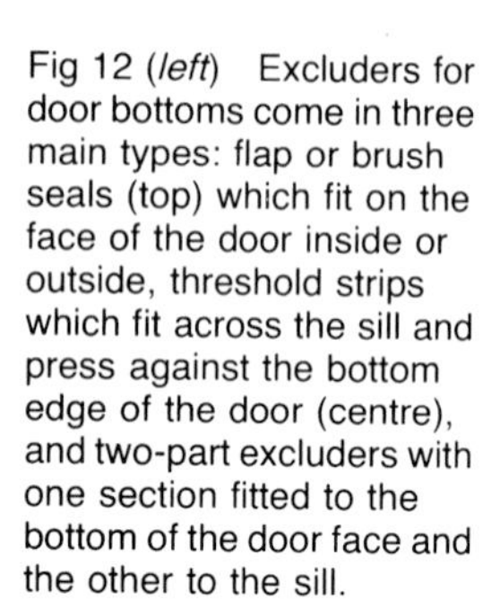

Fig 12 (*left*) Excluders for door bottoms come in three main types: flap or brush seals (top) which fit on the face of the door inside or outside, threshold strips which fit across the sill and press against the bottom edge of the door (centre), and two-part excluders with one section fitted to the bottom of the door face and the other to the sill.

Fig 13 (*top right*) **1** Flexible plastic sheeting. **2** Rigid plastic sheeting. **3** Framed panes secured with small turn-button clips. **4** Hinged types can be held closed with a clip on the side opposite to the hinges. **5** Two framed panes mounted in top and bottom tracks. **6** Sealed-unit double glazing.

Fig 14 (*bottom right*) Excluders for the sides of door frames and for windows come in a wider range of types. The cheapest are self-adhesive rubber or foam strips, which fit around the rebate and are compressed as the door or window closes. More expensive and more durable are the metal or nylon V-seals, sold in coils and pinned or stuck in place. The best but dearest are mouldings with brush or tube inserts which press against the closing face of the door or window when it is closed.

Double Glazing

The last category of insulation is double glazing, which is used to cut down on heat losses through the glazed areas of windows and glass doors. Heat losses here depend on the overall ratio of window to exterior wall, but are commonly 10 per cent of the total heat loss from the building. Double glazing can halve that, and the simplest way of installing it is to fit inner panes of plastic or glass over the existing windows – known as secondary glazing – so the film of still air trapped between them can act as an insulator. There is a wide range of DIY double glazing systems on the market, all available as kits which you make up to suit your window sizes.

The simplest system (1) consists of a thin flexible-plastic film held over the window with double-sided adhesive tape. It is easy to put up in autumn and cheap enough to remove and throw away in summer. Slightly more substantial types have rigid plastic panels held in place with magnetic strips (2). Glass panes need stronger support because of their extra weight. Framed panes can be held in place with clips (3), hinges (4) or sliding track (5).

As an alternative to secondary glazing, the existing panes can be replaced by sealed unit double glazing (6) (see pages 80–83 for more details).

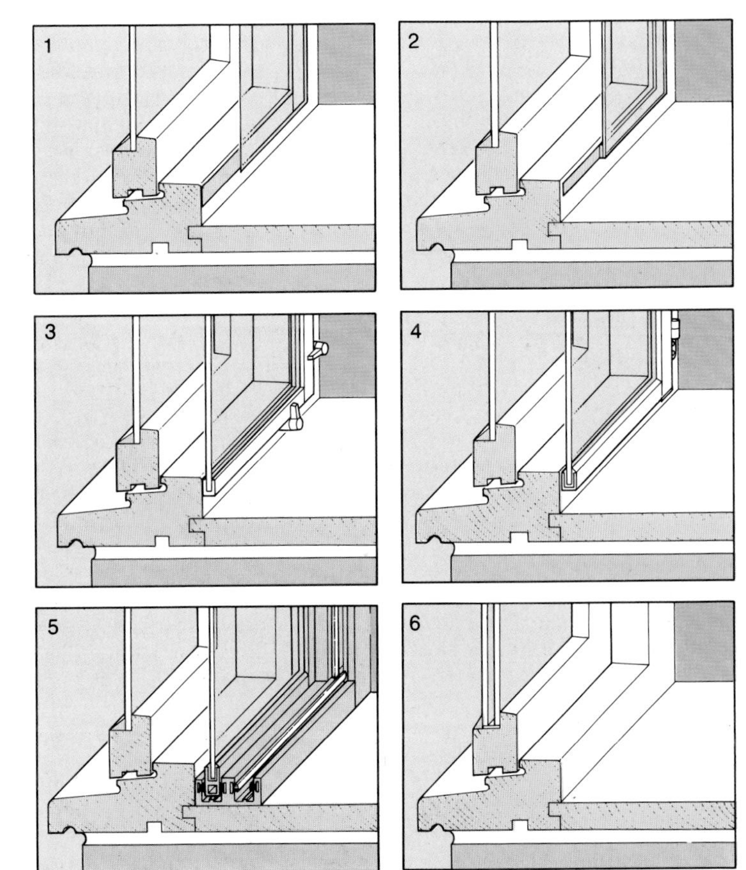

Shopping and Safety

You are likely to need quite a range of different materials to tackle repairs to your roofs and chimneys, keep the rainwater system in good order, make sure the exterior walls are weatherproof and attend to the various ways of insulating your home and its services. Where you shop for them depends on the scale and nature of the job. For a simple repair you may need to look no further than your local hardware store but, for more extensive projects, you will make considerable savings by shopping around for materials. Here are the places to try.

Local DIY Shops

The typical independent high street DIY shop usually stocks things like fillers, mastic, putty, small bags of dry ready-mixed mortar, flashing and other repair tapes, plus small sizes of products like liquid damp-proofers and possibly a small range of draught excluders. You will probably be offered a choice of just one or two brands. You should also be able to buy any general-purpose tools you need and do not own.

Verdict Fine for small jobs if they have what you need and the convenience outweighs the disadvantage of highish prices. Generally good at offering helpful advice, and a useful source of contact for local contractors for large-scale jobs.

DIY Superstores

The major national chains all offer a good range of materials for repairing or resurfacing flat roofs, including roofing felt, glazing sheets and so on, but rarely stock things like tiles or slates. They are also strong on loft and pipe insulation materials, with several brands to choose from, and most stock at least one brand of gutters and downpipes. All sell ladders and steps, and some operate hire sections which can be useful for access equipment and some specialist tools.

Verdict Good range of relevant products with the one exception of pure building supplies, usually at reasonable prices, but unlikely to offer much technical advice.

Builders' Merchants

A good source of supply for 'heavy' building materials like tiles (so long as they stock the make you want to match), slates and roofing felt, insulation (especially 'trade' materials like batts, plasterboard and polystyrene sheets), and for large quantities of things like liquid damp-proofers and masonry sealers. Some have 'retail' counters designed to cater for the non-trade customer.

Verdict Reasonable selection of goods at reasonable prices. Good for large-scale projects.

Roofing Specialists

Your best source of supply for slates, tiles of all types flat-roofing materials, flashings and other associated products such as eaves ventilators, plus specialist tools. Many also stock rainwater systems, and can usually carry out roofing work that is beyond the ability of the average do-it-yourselfer.

Verdict All the roofing products you are likely to need under one roof, at generally keen 'trade' prices. Good for technical advice.

Other Specialists

Most high streets contain at least one double glazing shop, mainly offering replacement windows and patio and front doors. If that is what you want, shop around and compare prices and specifications carefully before committing yourself. Personal recommendation is the safest guide to getting a good job at a keen price. Choosing a firm that is a member of a relevant trade organization is also a useful safeguard if things go wrong (see page 94 for more details).

Insulation contractors are the other specialists you may want to call in. Some are local firms, others branches of larger outfits operating nationally. Again, go by personal recommendation and look for membership of a relevant trade association.

Verdict Choose with care.

SAFETY

The main danger with roofing and outdoor weatherproofing jobs is of falling, either from access equipment or after climbing on to roof surfaces. The safe use of access equipment is covered in detail on pages 20–24; read these carefully before carrying out any high-level repair work. NEVER climb directly on to pitched roofs, and take great care near the edges of flat roofs.

Other general safety points to watch out for include:

- minor hand injuries – tiles and bricks have rough surfaces, so wear stout gloves when handling them
- chemicals – liquid damp-proofers and the like – can irritate skin and eyes, so wear gloves and safety goggles when applying them to guard against splashes
- insulation – some fibrous materials can cause severe skin irritation, so wear gloves, a face mask and long-sleeved clothing when handling them

Access Equipment

You will need a wide range of access equipment to carry out the various jobs in this book, especially as far as roof repairs are concerned. For safety's sake, always use the correct equipment for the job, and never try to improvise; one fall could be your last.

The basic item for access to the upper parts of your house walls, the eaves and the roof itself, is a ladder. An aluminium extension ladder is ideal, since it is fairly light and easy to move around, reasonably sturdy to climb and very durable. If you are buying one, choose a size long enough for the top to reach to a point about 1m (3ft) above eaves-level with an overlap of four rungs – never less – between the extended section. A triple ladder is often more useful than a double because you can separate the sections for access to different heights, and you can even use one of the sections as a roof ladder if you add special hooks to one end (see page 24). Alternatively, you can always hire both extension and roof ladders as and when you need them.

Roof ladders have a large C-shaped hook at one end, which is designed to lock over the ridge of the roof. Wheels on the opposite side of the ladder allow you to push it up the roof and into position. Make sure it is long enough to reach right down to the eaves, so you can climb on to it easily from your extension ladder.

If you find that working from ladders is uncomfortable, or you do not have a very good head for heights, you may prefer to hire an access tower. This is a slot-together structure which provides a stable working platform at whatever level you require, with the added safety feature of a handrail all round. Towers come in a range of heights, so make sure you order enough components to reach the height you want. Such towers can also be used to provide access to a variety of awkward-to-reach spots (see pages 22–23).

For low-level access – up to around the tops of ground-floor windows – you can either use steps or set up trestles to support scaffold boards in front of the area you are working on.

CHECK

- that any hired access equipment is in good condition, with ladders having secure rungs and straight stiles (sides) and access tower components undamaged and free from rust

Fig 15 (*left*) An extension ladder is a must for many outside repair and maintenance jobs. For large-scale projects, hiring a slot-together access tower is worth considering, especially if you find ladders uncomfortable to work from or you dislike heights.

Using Ladders Safely

Falls from ladders are the single biggest cause of death and injury in accidents around the home, and they are almost always caused by a combination of carelessness and stupidity on behalf of the person using them (modern alloy ladders are almost indestructible, so ladder failure is virtually unheard-of). For this reason, learning to use ladders safely is an essential skill to acquire.

Setting up the Ladder

The safest way of setting a ladder in position is to carry it to where you plan to erect it and lay it flat on the ground with its foot against the base of the wall. Then lift the other end above your head with both hands, and walk under it towards the wall, moving your hands from rung to rung until the ladder is vertical.

Next, move the foot of the ladder out from the base of the wall until it is at a distance from the wall equal to one quarter of the height of the top of the ladder. This ensures that it is standing at the correct angle, neither too steep nor too shallow. Check that the ladder is vertical, when viewed from straight on to the wall.

If you are using a single section of ladder, or a short and unextended ladder, you can instead simply carry the ladder in the vertical position to where is is needed and set it straight into position. Beware of overhead obstructions as you carry it around the house.

Extending the Ladder

If the ladder has two or three sections, it is often easier to pull them out to the required height while the ladder is still on the ground. The foot of each section has two clips which fit over a rung in the section beneath; simply unhook these, pull the section out and then hook the clips over the appropriate rung. Always ensure that there is a four-rung overlap between adjacent sections.

When the ladder is in place, you can raise or lower it slightly to get precisely the height you want, but do this only if you can reach the bottom of the upper section while standing on the ground.

Some hired extension ladders are much sturdier and heavier than DIY versions, and have a rope-operated pulley system to make extending the ladder easier.

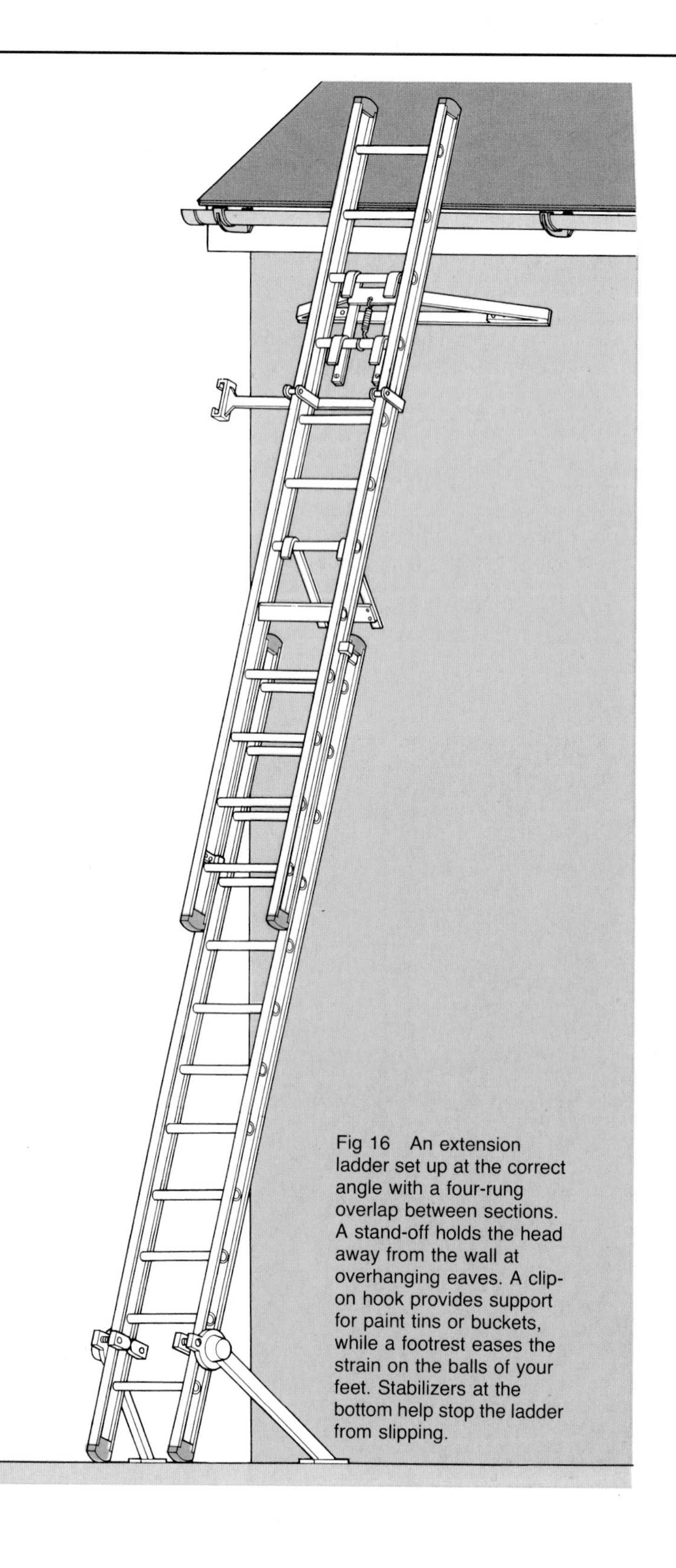

Fig 16 An extension ladder set up at the correct angle with a four-rung overlap between sections. A stand-off holds the head away from the wall at overhanging eaves. A clip-on hook provides support for paint tins or buckets, while a footrest eases the strain on the balls of your feet. Stabilizers at the bottom help stop the ladder from slipping.

Using Ladders Safely

Securing the Ladder

This is the most important part of the operation. To be safe to climb, the ladder must be standing level, square to the wall, and should always be secured to stop it slipping, either outwards at the base or sideways at the top.

On hard, level surfaces such as a path or patio, set the foot of the ladder on some sacking and secure it with a heavy bag, for example, a sack of soil or a bag of cement. On soft ground, do not just tread the foot of the ladder into the soil, because as you climb, one stile may sink faster than the other and topple you sideways off the ladder. Instead, nail a batten across a board offcut, place this on the soil and set the ladder up on it. As an added precaution, tie the foot of the ladder to stout pegs driven into the soil.

Next, you should secure the head of the ladder to the building. If a window is nearby, it may be possible to tie a rope around part of the window frame, or to set a baulk of timber across the inside of the opening and tie the ladder to that. Never trust downpipes to provide a fixing point, as they may pull away from the wall.

The best solution is to take the time to position stout screw-eyes in holes drilled in the house wall at intervals equal to your span with outstretched arms – roughly 1.6m (just over 5ft) is typical. Then you can easily tie the ladder to at least one eye at any point around the house walls. Use all-in-one expanding anchors with a screw-eye fitting, tightening the eye bolt into the anchor after inserting it in its hole by turning it with a screwdriver blade through the eye. Make sure each anchor is set into the centre of a brick, not into the mortar in between, so it cannot pull out under load.

Ladder Accessories

You can fit a variety of accessories to your ladder to make it safer or easier to use. These include a stand-off to hold the head of the ladder clear of the wall for use near overhanging eaves, foot rests, tool racks and stabilizers.

TIP 1
Always wear shoes with thick, rigid soles and a positive grip when working on a ladder. The rungs will soon cut through soft, thin soles and make your feet ache.

TIP 2
When climbing a ladder, keep your arms almost straight so you are leaning slightly backwards, as this is much more comfortable than clutching the rungs to your chest. When you reach the height you want, keep your hips within the line of the stiles at all times so you are not tempted to over-reach to either side of the ladder.

TIP 3
Keep hold of one stile of the ladder at all times if possible. Do not carry heavy items up with you; either haul them up after you on a rope, or get someone to pass them to you through a window.

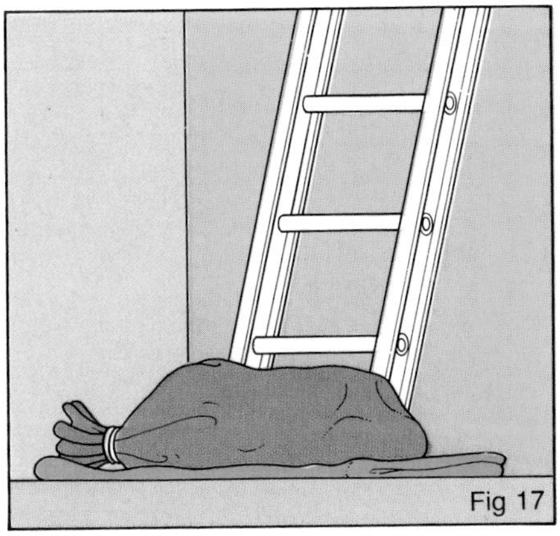

Fig 17 On solid ground, set the foot of the ladder on sacking and weight it with a sack of soil.

Fig 18 On soft ground, set the foot on a board with a batten nailed to it, and tie the ladder to two stout stakes.

Fig 19 At the top, tie the ladder to a window frame or to a baulk of timber set inside the window opening.

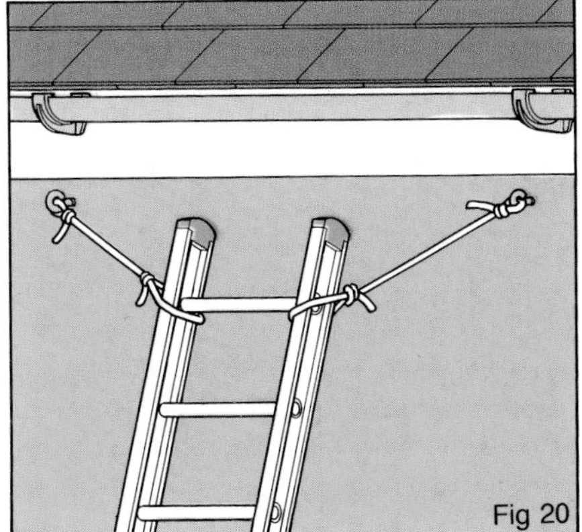

Fig 20 Alternatively, position screw-eyes at high level all around the house walls and tie the ladder to these.

Using Access Towers

Access towers are an excellent alternative to ladders if you need access to larger sections of the house exterior or you are tackling a long and fiddly job which would be uncomfortable to carry out from a ladder. They are also safer than ladders as long as they are properly erected, and are ideal for people who have a poor head for heights because they provide a large stable working platform with a handrail all round the top.

Unlike ladders, few people own an access tower. Unless you are carrying out lengthy and regular work on your house, it is far more economical to hire one when you need it from a local plant hire firm.

Tower Components

The tower consists of H-shaped galvanized steel or aluminium alloy frames which are slotted together in alternating parallel pairs until the desired working height is reached. Diagonal braces stiffen the structure at intervals, while square feet spread the load at ground level. These can be adjusted to get the tower level on uneven ground, and can be replaced by lockable castors to make a mobile tower if required. Outriggers can be fitted to improve the stability of high towers. Most are 1.2m (4ft) square, but half-width frames are also available for building a smaller rectangular tower in confined spaces.

At the top of the tower, handrails are added to link the last pair of frames, and a platform of flat boards is laid. Toe boards prevent tools from falling or being kicked off the platform. The maximum platform height with this type of lightweight tower is usually 4.8m (16ft), giving a working height of about 6.5m (22ft) from the platform, which is enough for most homes. Taller industrial-quality towers are available if necessary, with larger bases to increase stability.

Some alloy towers have internal ladders and intermediate platforms to make it easier to climb the tower and carry materials up to the working platform.

The components of these towers can also be used to erect low-level working platforms for use indoors or outside, and extra components can be added to allow the construction of platforms on stairwells or around chimney-stacks. They can even be used in conjunction with staging to form a platform spanning obstructions or door openings (see opposite).

Erecting the Tower

Decide on the site for the tower, and fit the feet or castors to the lowest pair of frames. If you are using castors, make sure they are locked. Then stand the frames up and link them with the first diagonal brace. Use a spirit level to check that the tower is standing squarely, and add the next pair of frames at right angles to the first, followed by another brace fitted across the opposite diagonal to the first one. Then climb up inside the tower and get a helper to pass you up the next two frames; fit them and add another brace. When you reach the height at which you want the platform,

Fig 21 A typical DIY access tower consists of H-shaped frames built up in pairs to the desired height. Diagonal braces stiffen the structure, while outriggers improve its stability. At the top, boards form a work platform and handrails improve safety and security.

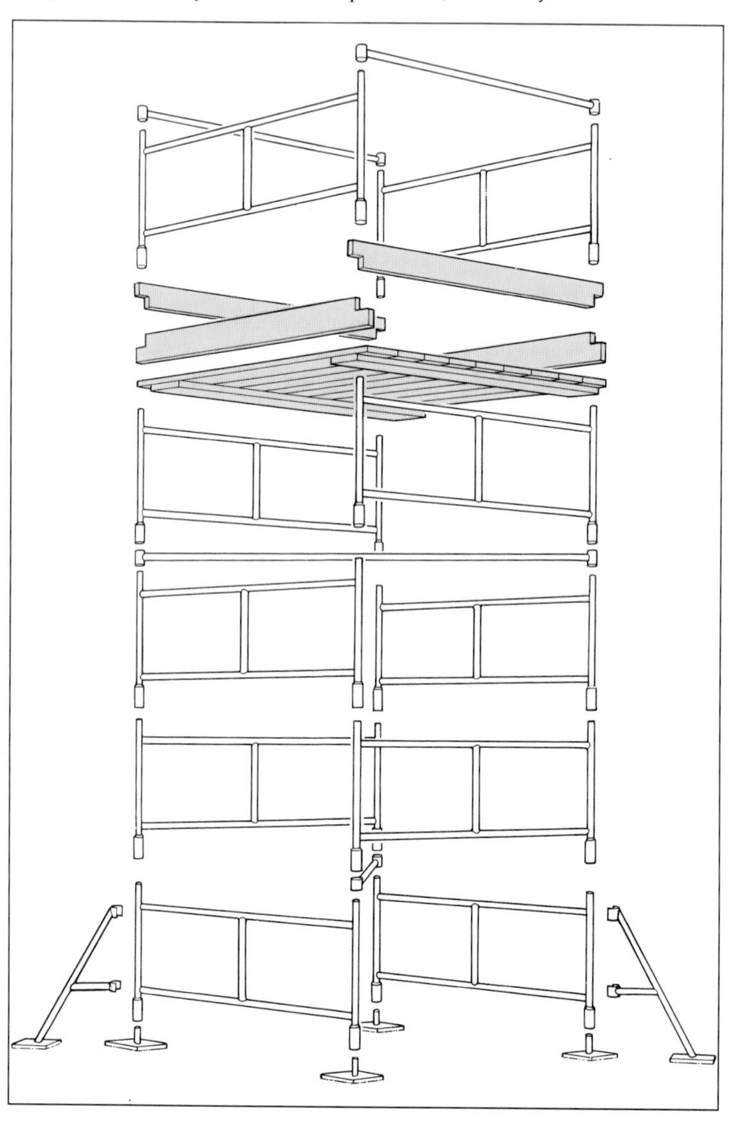

add on the last pair of frames and lay the platform and toe boards. Finally, fit the handrails to link the last pair of frames.

If the tower is on a level hard surface and is fitted with castors, these can be unlocked to allow the tower to be moved along as necessary. However, it should never be moved when anyone is on it. On soft ground, fixed feet must always be used and they should be set on bearer boards to help spread the load and stop the tower sinking into the soil. Adjustable types allow you to get the tower standing absolutely square on slight slopes.

Lift a couple of the platform boards for access on to and off the platform; remember that you must always climb up inside the tower, never on the outside. For towers over about 4m (13ft) high, you should tie the framework to the building to prevent any risk of it toppling over. Do this by roping it to a window frame, to a baulk of timber inside a window opening or to stout screw-eyes fixed to the house walls, as described on pages 20–21 for securing ladders.

Unusual Access Problems

You can use the components of access towers to cope with a wide variety of awkward access problems. For example, you can create a cantilevered structure to gain access to walls or eaves above projections such as bay windows or porches, where it is not possible to set a ladder up at the correct angle. Similarly, the tower can be built out sideways over a flat roof, both to provide safe access on to the roof and to act as a working platform so you can reach the walls or eaves above it.

Twin towers with lengths of staging between them can be used to bridge obstacles, or to leave free access to a doorway.

You can also use the tower components to form a safe working platform around a chimney-stack, either improvising supports to protect the roof surface from damage or hiring special chimney-stack access kits. These are far safer than trying to work on a stack while balanced on a roof ladder, and provide somewhere to store materials until they can be lowered to the ground.

CHECK

- that you have all the components you need when a hired tower is delivered; missing feet or braces could make it impossible to build the tower correctly
- that handrails and toe boards are supplied, and that you fit them
- that the tower is built to a true vertical, using adjustable feet on uneven ground
- that castors are locked before anyone climbs the tower
- that children are not allowed to play on or near the tower

TIP

Oil the frame sockets lightly as you assemble the tower, to make it easier to dismantle after use.

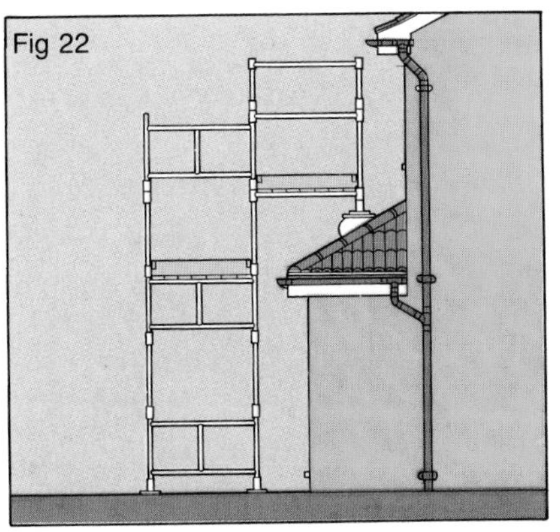

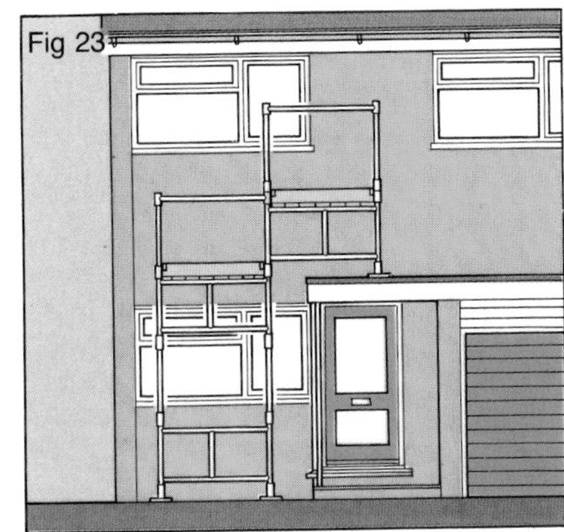

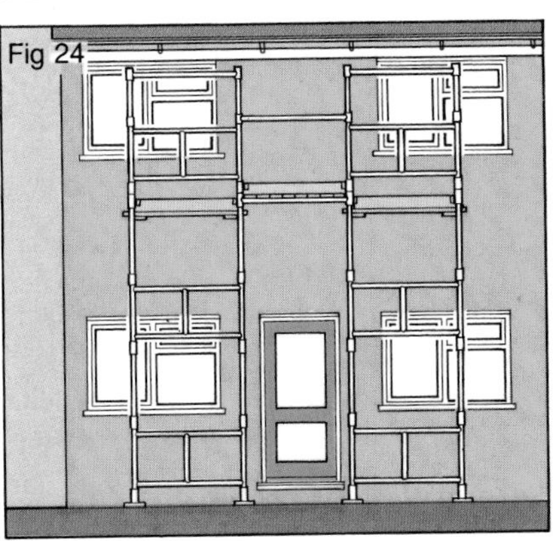

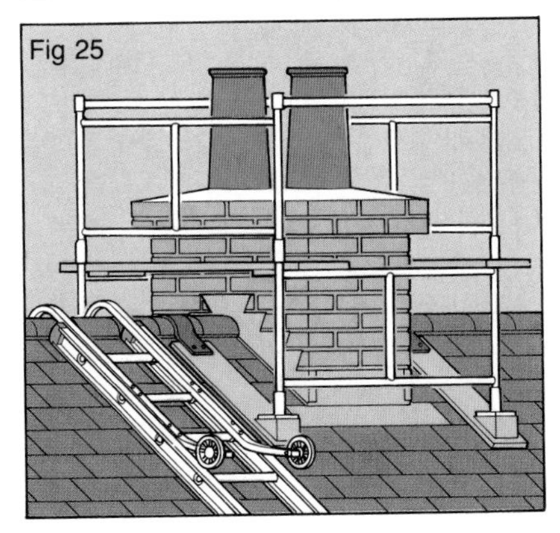

Fig 22 Cantilever the tower over projections such as porches, resting the frames on a board laid on sandbags to protect the roof surface.

Fig 23 Build out sideways over flat roofs for safe access to the roof and the walls above.

Fig 24 Build twin towers linked with staging to span openings such as doorways.

Fig 25 Use kit components or special stack kits to build a safe working platform around chimney-stacks.

Working on the Roof

If your house has a pitched roof, you may need access to it to replace damaged or missing tiles, to rebed ridge-tiles or to gain access to chimney-stacks for repair or maintenance work. The only safe way of doing this is via a roof ladder; you should never attempt to emulate professional roofers and climb directly on to a tiled or slated roof, since one slip would almost certainly send you hurtling to the ground with serious or fatal injuries.

Roof ladders have a large C-shaped hook that fits over the apex of the roof, and a pair of small wheels on the opposite side of the ladder to allow you to push it up the roof slope from the eaves. When the wheels reach the ridge, you simply turn the ladder over so the hook engages over the ridge. Care needs to be taken not to dislodge the ridge-tiles, and if you find any that appear to be loose you should rebed them in fresh mortar immediately.

You can hire roof ladders from local plant hire firms, or convert a section of your own extension ladder for use as a roof ladder by bolting on a pair of special wheeled roof hooks. The ladder you use should be long enough to reach right down to the eaves. Common sizes for hired ladders are 5, 6 or 7m (16, 19 or 22ft) long. It does not matter if the ladder is a little too long for the roof slope, so long as you take care not to put any weight on the projecting section as you climb on or off the roof. If you do, the hook could disengage and allow the ladder to tip up off the roof, taking you with it.

To provide safe access for positioning and climbing on to the roof ladder, you need to erect your extension ladder so that it projects above the eaves by about 1m (3ft). This is essential to allow you to reach the end of the roof ladder safely, with something to hold as you climb on to it. Rope the top of the extension ladder to the house wall and then tie the bottom of the roof ladder to the extension ladder so nothing can move.

A word of warning: if you are unhappy working at heights, leave work on your roof to a professional roofing contractor rather than risk an accident.

TIP

If you need to move the ladder across the roof, do not try to slide it sideways. Instead, simply turn it over and over until you reach the desired position, then re-engage the roof hook.

CHECK

- that the roof hook is right over the ridge before turning the ladder over
- that ridge-tiles are securely bedded
- that the foot of the roof ladder is tied to the top of the extension ladder before climbing on to it
- that the extension ladder projects above the roof by at least 1m (3ft) and that it is tied to the house wall

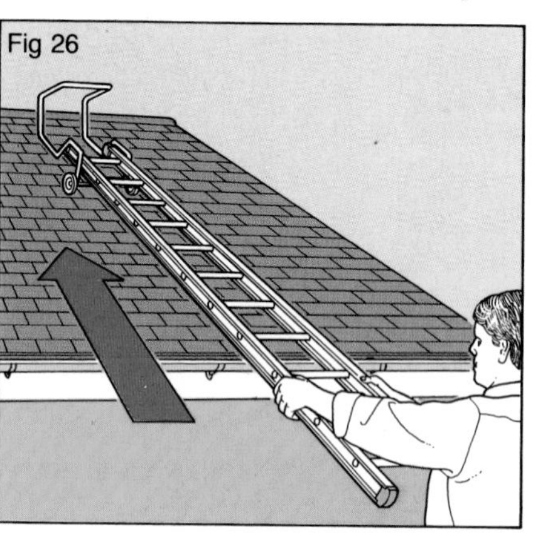

Fig 26 Haul the ladder up to eaves-level, then raise it on to the roof slope with the wheels downwards. Push it carefully up the roof slope so you do not dislodge or damage tiles or slates.

Fig 27 When the wheels reach the ridge, turn the ladder over to engage the roof hook.

Fig 28 Tie the bottom of the roof ladder to the top of your extension ladder, and tie that to the house wall.

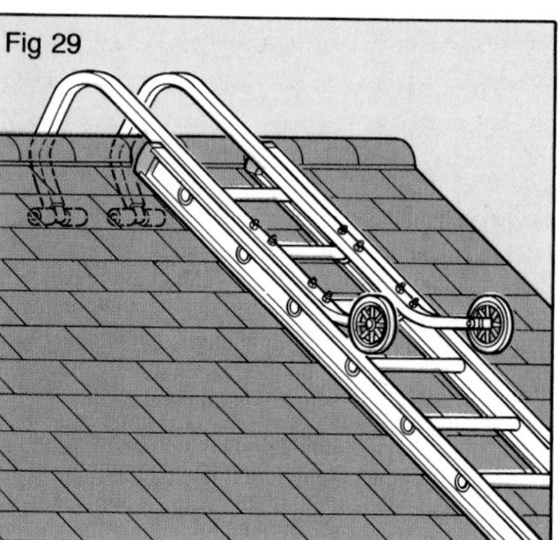

Fig 29 As an alternative to hiring a roof ladder, fit a pair of roof hooks to a section of your extension ladder.

ROOFING JOBS

There is as much variety in the make-up of roofs on domestic property as there is in any other feature – walls, windows, chimney-stacks and so on. All have evolved according to regional traditions and the availability of suitable materials, although nowadays much of the variety has gone out of new buildings thanks to the standardization and uniformity of modern building materials. On the other hand, it is true to say that modern products and fixing methods now offer excellent performance and easier installation than traditional types, and can still produce an elegant and visually attractive roof.

The one backward step in roof technology is without doubt the felted flat roof, which has topped many a budget-priced home extension during the last 30 or 40 years and has caused more performance problems than almost any other building element.

Most homes in this country have pitched roofs, with the sloping surfaces covered with either tiles or slates. The simplest form of pitched roof is known as *a gable* roof, and consists of two sloping surfaces meeting at a ridge, with the A-shaped ends filled in with triangles of brickwork, which are the gables. Slightly more complex in terms of the framework needed to support it is the *hipped* roof, which has triangular sloping roof surfaces instead of vertical gables at each end of the roof. The intersections between adjacent sections of the roof slope are known as the hips, and give the roof its name.

Both these common roof types are often used on houses with L-shaped or T-shaped floor plans, as well as those that are just squares or rectangles. The adjacent roof sections meet at internal angles called valleys, which may be formed of interlaced tiles or slates or may be lined with lead or zinc sheeting, known as a valley gutter. At a gable end, the edges of the roof surfaces are called verges, while the bottom edge of the roof surface is the eaves. Where a roof slope meets a vertical surface such as a chimney-stack or, in the case of a lean-to roof, the house wall, the junction is known as an abutment.

At the eaves, the rafter ends are concealed by a vertical fascia board and, where the eaves overhang the walls, the underside of the overhang is boxed in with a soffit. Overhanging verges at gable ends are usually finished off with a pair of barge-boards, again with soffit boards behind them.

Fig 30 Removing a broken tile using a trowel (*see* opposite).

Other less common roof types are the *gambrel*, the *mansard* and the *inverted* or *butterfly* roof. Gambrel and mansard roofs have each slope formed by two sets of rafters; the lower set are very steep, while the upper set are at a shallower pitch, allowing rooms to be formed within the roof space while saving on the use of more expensive walling materials. Gambrel roofs have gabled ends; mansards are hipped.

Inverted roofs are gabled roofs in reverse, with high-level eaves at each side of the roof and the two slopes meeting in the centre of the building in a low-level valley. They are often found on Georgian and some Victorian terraced houses, but are seldom built nowadays.

Lastly, the flat roof – in fact built with a gentle slope to encourage rainwater to run off – has a decking of timber or man-made board, which is usually covered with either a three-layer system of roofing felt or a layer of mastic asphalt. The lower end of the roof slope finishes at the eaves in the same way as a pitched roof, while the sides of the roof either have verges or built-up parapet walls.

Replacing Tiles

There are several reasons why you may need to replace tiles on your roof. The nibs holding individual tiles over their supporting battens may have broken, allowing the tile to slip out of place. The tile itself may have been cracked, usually by someone climbing on the roof surface to carry out repair work. Worst of all, tiles may have been lifted and literally flung from the roof by high winds.

On roofs with plain tiles, each tile is hooked over a batten nailed across the rafters, and these battens are closely spaced to allow each tile to overlap roughly half the length of the tile below it (Fig 31). Tiles in every third or fourth row are usually nailed to their batten to help prevent high winds lifting the tiles. This is known as double-lap tiling.

On roofs with interlocking tiles, the battens are more widely spaced and each tile overlaps the one below it by only 65–75mm (2½–3in). The tiles overlap slightly and also interlock with their neighbours at their vertical edges. This is known as single-lap tiling.

What to do

If you have tiles missing from your roof, you need to inspect the fault and assess the scale of the repair. This means setting up a ladder secured to the eaves, and, if necessary, a roof ladder as well to allow you to climb safely on to the roof surface (see pages 20–21 and 24).

Use timber wedges to force up surrounding whole or undamaged tiles so you can lift out any broken pieces. If the tiles appear to be nailed, use a tool called a slater's ripper (available from hire shops) to cut through the nails so you can remove the tile. If several tiles are damaged or missing, strip the area course by course.

To obtain matching replacements, take a whole tile with you to your supplier; if you can obtain weathered second-hand tiles, so much the better. To replace a single tile, simply slide it into place between the wedges until its nib engages over the batten, and remove the wedges to allow the adjacent tiles to fall back into position. For larger areas, replace the tiles

What you need:

- access equipment
- wooden wedges and hammer
- slater's ripper
- replacement tile(s)
- non-ferrous nails or fixing clips
- mortar for verge and ridge-tiles
- bricklayer's trowel

CHECK

- that surrounding tiles are undamaged, by tugging them to ensure that their nibs are intact and hooked over the tiling battens
- that tiling battens are free from rot; replace any damaged sections before fitting the new tiles

TIP

If plastic clips have been used to secure the edges of interlocking tiles to the battens, simply cut through them with a sharp knife to free the tile edge. Fit new clips if you are replacing an area of tiles.

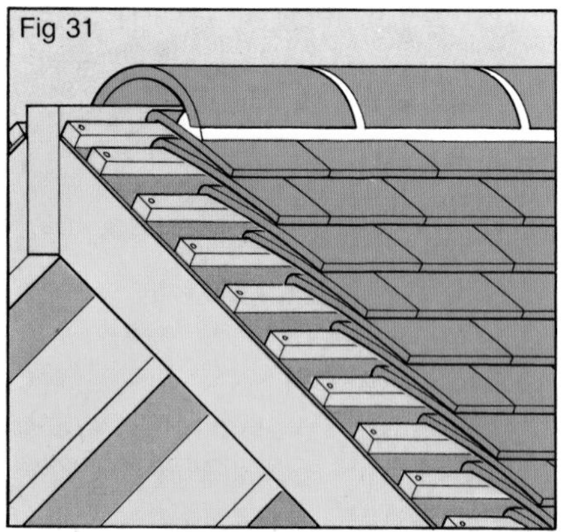

Fig 31 Plain tiles are laid in double-lap fashion over closely spaced tiling battens.

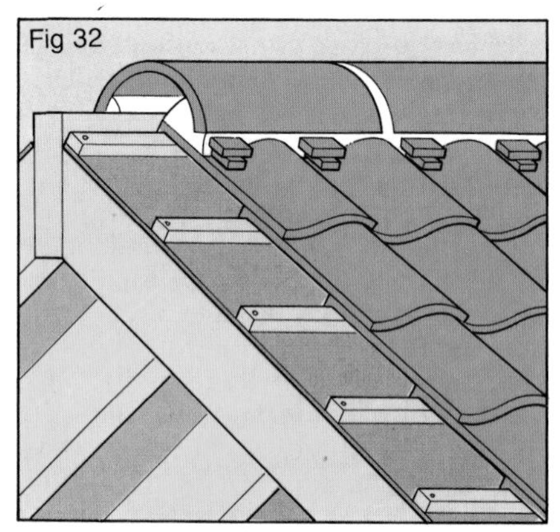

Fig 32 Interlocking tiles are laid in single-lap fashion, with battens spaced further apart.

Fig 33 To remove a damaged tile, start by wedging up the tiles in the course above.

Fig 34 Then use a trowel to lift the nibs at the head of the tile off the tiling batten so you can remove the broken tile.

working up from the lowest course removed. Nail tiles at every third or fourth course, and add side clips if you are replacing an area of single-lap interlocking tiles.

If plain tiles are missing from the roof verge, you will need wide tile-and-a-half verge tiles to complete every other course. Set them in place on their battens, nailing every course, and bed the verge edge of the tile on mortar. With interlocking tiles, you can use the same tiles at the verge as elsewhere on the roof.

Replacing Ridge-Tiles

A tiled roof is finished off with a row of shaped tiles running along the ridge (and sometimes down the hips as well on hipped roofs). These are bedded on mortar and if this cracks the bond may be lost, allowing wind to lift and dislodge the tile. If the tile is still intact and on the roof, remove the old mortar from its underside and the bedding mortar from the roof slope. Then bed it back on fresh mortar, pointing the joints between it and the other tiles.

Fig 35 (*above*) Move the wedges up one course at a time to replace tiles over several courses.

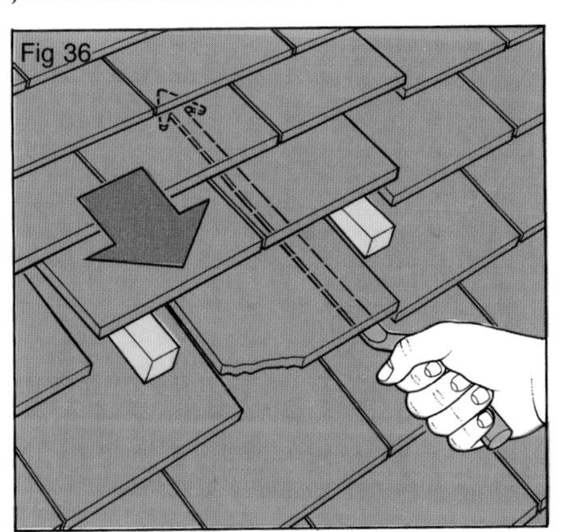

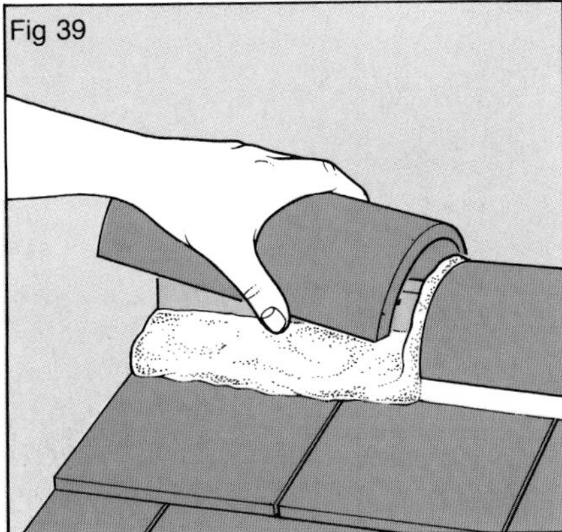

Fig 36 If the damaged tile is in a nailed course, use a slater's ripper to cut through the fixing nails. Slide the tool up past the nail, hook the cutting jaw round it and pull sharply downwards.

Fig 37 Slide the replacement tile in, hooking its nibs over the batten. Then remove the wedges.

Fig 38 Lift a loose ridge-tile and chip away the old mortar bed.

Fig 39 Replace the tile in a bed of fresh mortar, ensuring that there is a neat joint between it and neighbouring tiles.

Replacing Slates

The commonest fault affecting slated roofs is nail sickness – an aptly named defect. It means that the nails securing each course of slates to their batten have rusted away, allowing the slate to slip out of position. Slate is also prone to flaking and delamination as, with time, rainwater starts to penetrate the layered structure and then freezes, splitting the stone. Lastly, it can of course suffer physical damage, resulting in slates cracking and unsecured sections falling out.

As with plain-tiled roofs, slated roofs are laid in double-lap fashion, with each slate overlapping the one below it by just over half its length. The reason for this is that slates, unlike tiles, are by their nature flat, and capillary action can cause rainwater that penetrates the vertical joints to seep sideways between the slates. The double lap stops further penetration.

Slates are usually fixed to the battens beneath by nails driven through pre-drilled holes either at the head or half-way down their length; short eaves and ridge-slates always have the holes near the top edge.

What to do

If a slate is cracked or otherwise damaged but remains fixed in place, you need a slater's ripper to cut through the fixing nails and free the slate. Slide it up beneath the affected slate and hook its jaws around the nail. Then give the handle a sharp tug downwards. Repeat the process for the other nail. Then slide the old slate down the roof slope and remove it.

Since you cannot wedge up the course above without breaking the slates, you will be unable to gain access to the line of the batten to which the removed slate was nailed. Therefore, to hold it in place, the answer is to fix a strip of lead or zinc to the roof slope with a nail driven between the two slates that lie beneath it. Position the nail in line with the fixing nails holding the exposed slates, so you can be sure it is driven into the batten beneath. Then slide the replacement slate into position over the metal strip, aligning its bottom edge with its neighbours, and bend up the strip to hold the new slate securely in place.

What you need:
- access equipment
- slater's ripper
- replacement slate
- non-ferrous nails
- lead or zinc strip
- hammer

CHECK
- that any replacement slates are the same thickness as the originals. If you cannot get a replacement of the right size, buy a larger one and cut it down with a masonry cutting disc or ask your supplier to cut one to size for you

TIP
If you are replacing several slates which have enlarged nail holes, drill out new nail holes with a masonry drill bit

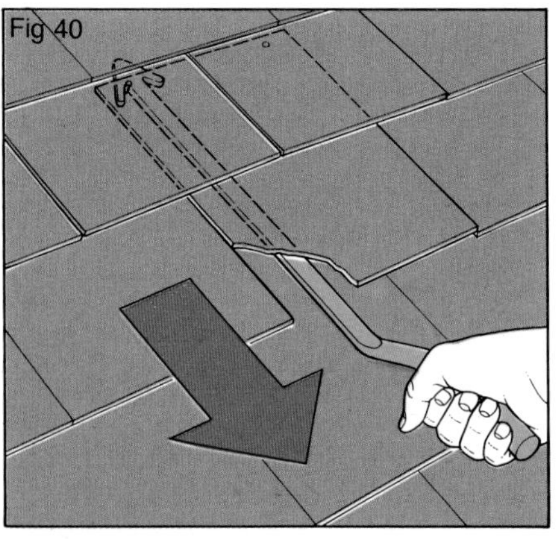

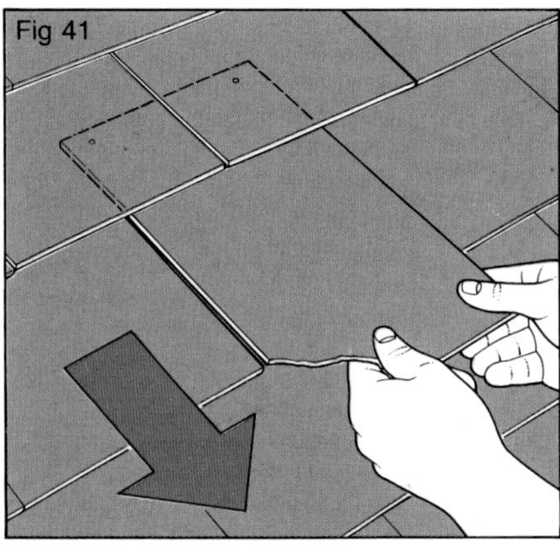

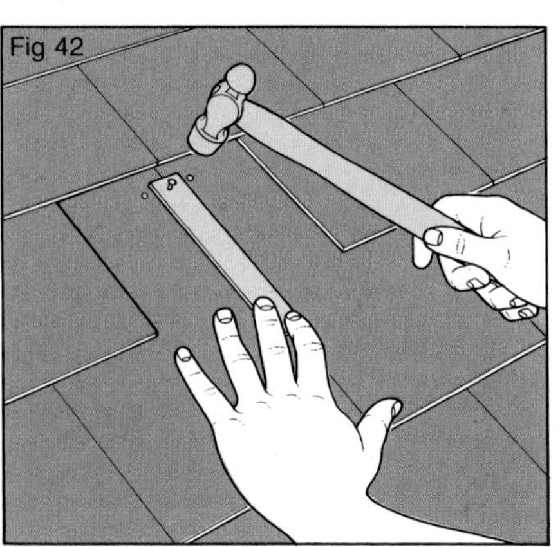

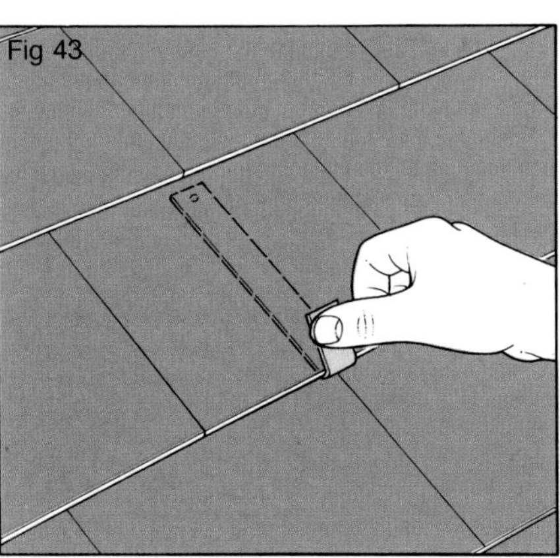

Fig 40 To free a damaged slate, slide a slater's ripper up beneath it and hook the cutting jaw around the fixing nail. Then tug the handle sharply downwards to cut through it.

Fig 41 Slide the damaged section of slate out down the roof slope.

Fig 42 Nail a strip of lead or zinc to the roof, driving the nail between the slates and in line with their fixing nails.

Fig 43 Slide the replacement slate into place and secure it by folding the end of the metal strip up over its lower edge.

Patching a Flat Roof

Flat roofs covered in roofing felt are a source of regular problems, especially if they are mistreated or poorly maintained. One of the commonest faults is caused by the gradual loss of the protective top layer of mineral chippings, which are designed to prevent heat from the sun blistering and cracking the felt. Another is the use of the roof surface as a work platform for access to an adjacent wall or pitched roof; this results in chippings being ground into the felt layers, which can eventually cause pinholes. This is enough to allow water to seep beneath the felt, causing further blistering and leading to rot in the decking and the supporting joists. Carelessly used ladders and steps can also cause localized damage to the felt layer and again result in water penetration.

Also prone to damage and deterioration are the raised side edges of flat roofs, known as welted aprons. Here there is no protective layer of stone chippings to ward off the sun's heat, and water penetrating here can then run back beneath the main felt layer.

What to do

If you spot (or cause) localized damage to the surface of a felt roof, attend to it immediately. If the top layer of felt is blistered or split, scrape away any remaining chippings, make two cuts over the area at right angles to each other and peel back the flaps. Leave the area open for a while to allow any trapped moisture to dry out, then play a hot-air gun over it while you trowel in a small quantity of bituminous mastic. Then bed the flaps down into the mastic, secure them with four galvanized clout nails and brush over the repair with flashing primer. When this is dry, cut a piece of self-adhesive flashing tape, peel off the backing paper and lay it over the primed area, warming it with your hot-air gun to improve its flexibility. Press it down firmly, then run a wallpaper-seam roller over it, to exclude air and ensure a good bond.

Use the same technique to repair damaged welted aprons, choosing tape wide enough to extend from the face of the verge batten, over it and on to the roof surface itself.

What you need:

- sharp handyman's knife
- bituminous mastic
- hot-air gun
- galvanized clout nails
- hammer
- flashing primer
- self-adhesive flashing tape
- wallpaper-seam roller

CHECK

- that moisture is thoroughly dried out before repairing blisters, or the sun's heat will cause another blister to appear
- that all exposed felt is covered with mineral chippings to protect it from the sun. Spread extra chippings over bare areas

TIP

To avoid causing further damage as you make the repair, lay scaffold boards or other board offcuts over the roof surface to spread your weight as you work.

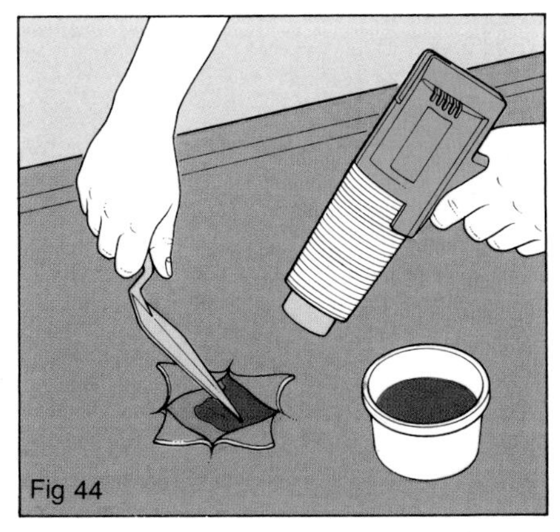

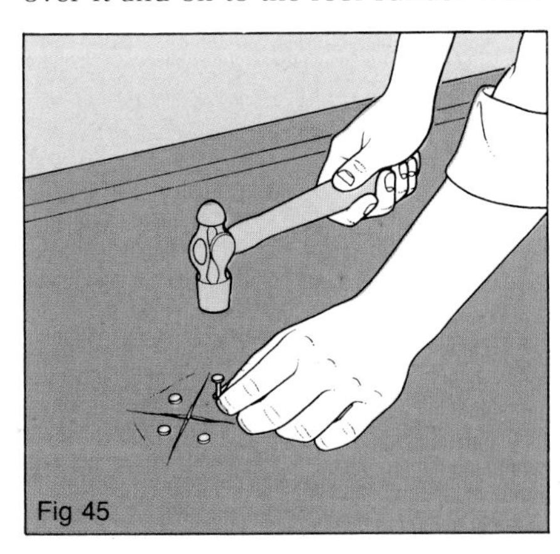

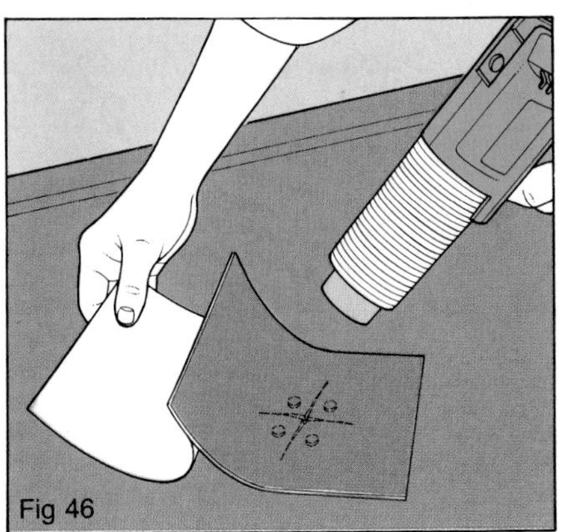

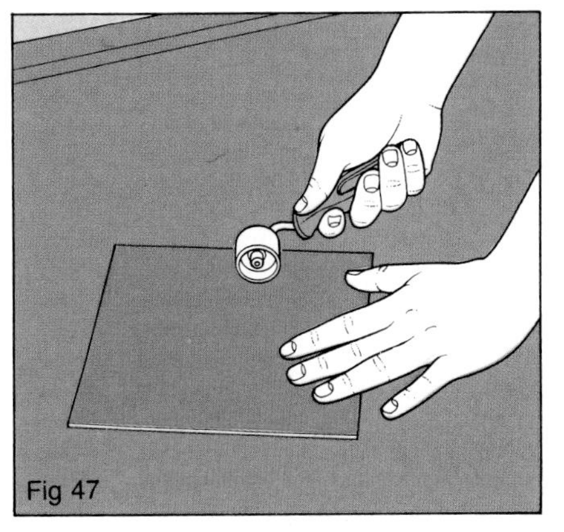

Fig 44 Cut blisters open with a sharp knife, dry out moisture and apply bituminous mastic.

Fig 45 Press the flaps back down and secure each with a galvanized clout nail. Then brush flashing primer over the repair to seal it.

Fig 46 Peel the backing paper off a piece of self-adhesive flashing tape and press it over the repair. Use heat from a hot-air gun to make it more flexible.

Fig 47 Finally use a seam roller to bed the patch firmly to the roof surface, and cover it with fresh chippings.

Resurfacing a Flat Roof

If a felted flat roof deteriorates to the point where water is actually staining the ceiling beneath, matters have reached a serious point and patching is not likely to offer a satisfactory solution. In any case, there is no guarantee that water will show up indoors directly below the source of the leak. Once it penetrates the felt covering, it can run down to the next joint in the decking, then along a joist, perhaps collecting within any insulation included in the roof structure before finally penetrating the plasterboard ceiling itself. Therefore, it is best to consider either coating the entire roof surface with a heavy-duty waterproofer, or stripping off all the felt and resurfacing the roof from scratch. The former is well within the capabilities of any do-it-yourselfer, but refelting is best left to a professional roofer, since the best results are obtained by bonding the felt layers together with hot bitumen, and this is definitely not a user-friendly material in non-expert hands. Also, tackling the job yourself and bodging it could be an expensive mistake.

What to do

The first step is to remove as many of the mineral chippings as possible from the roof surface. Most may long ago have blown off or washed into the gutters, but the remainder may have become stuck to the felt surface. Work on a small area (about 1sq m – 10sq ft) at a time, using your hot-air gun to soften the surface so that you can literally shovel the chippings off with a garden spade. Then sweep the entire surface thoroughly, and inspect it for obvious faults such as splits, cracks and tears. Stick any lifting areas down with mastic or roofing-felt adhesive.

If you are using a heavy-duty waterproofer, start at one edge of the roof and apply it in bands about 1m (3ft) wide. Leave it to dry, then apply a second coat in bands at right angles to the first, to ensure that any pin-holes and missed areas are well covered.

You can leave the surface as it is, but it is better to protect it from the sun by adding a thin layer of fine chippings or using solar-reflective paint (*see* TIP).

What you need:
- hot-air gun
- garden spade
- garden broom
- liquid waterproofer
- applicator broom (to be thrown away)
- reinforcing membrane
- sharp handyman's knife
- banister brush
- large paintbrush
- mineral chippings
- old boots and gloves (to be thrown away)

TIP
Cover the roof surface with solar-reflective paint rather than chippings if you intend to use it as an access platform for future maintenance work on nearby roofs or walls. Otherwise you will tread chippings down into the new roof surface and penetrate it again.

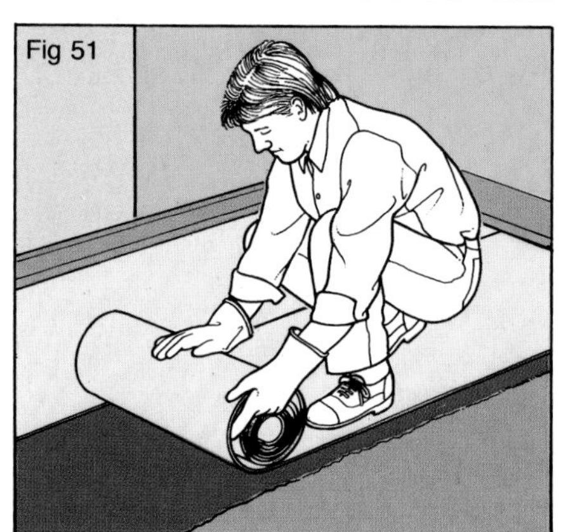

Fig 48 Remove chippings from the felted surface with a garden spade, using a hot-air gun to soften the felt and release any stuck-down stones.

Fig 49 Sweep all chippings and debris from the roof surface.

Fig 50 Brush on the first coat of waterproofer with a disposable broom, working in parallel strips across the roof. Unless you are using a membrane, apply the second coat at right angles to the first.

Fig 51 If you are using a membrane, unroll it while the waterproofer is still tacky, cutting it to length to run between the welted aprons at each side. Overlap lengths slightly.

Resurfacing a Flat Roof

An alternative to bitumen-based waterproofers is a urethane-based liquid rubber formulation. This comes with its own primer, which is thin enough to be applied to the prepared roof surface with a large paintbrush. This is followed by a generous application of the waterproofer, applied with a broom.

Using Reinforcing Membrane

Where the felt surface is seriously crazed, you will get better long-term results by incorporating a reinforcing membrane; this will help to resist further cracking. Prepare the roof as before, then brush on the first coat of waterproofer. While it is still tacky, unroll the membrane one length at a time, standing on it and following the roll across the roof as you work. Overlap parallel lengths by about 50mm (2in).

Next, scrub more waterproofer into the weave of the membrane, leave it to dry and finish off by applying a final topcoat over the whole surface. Complete the job by raking out a uniform layer of chippings.

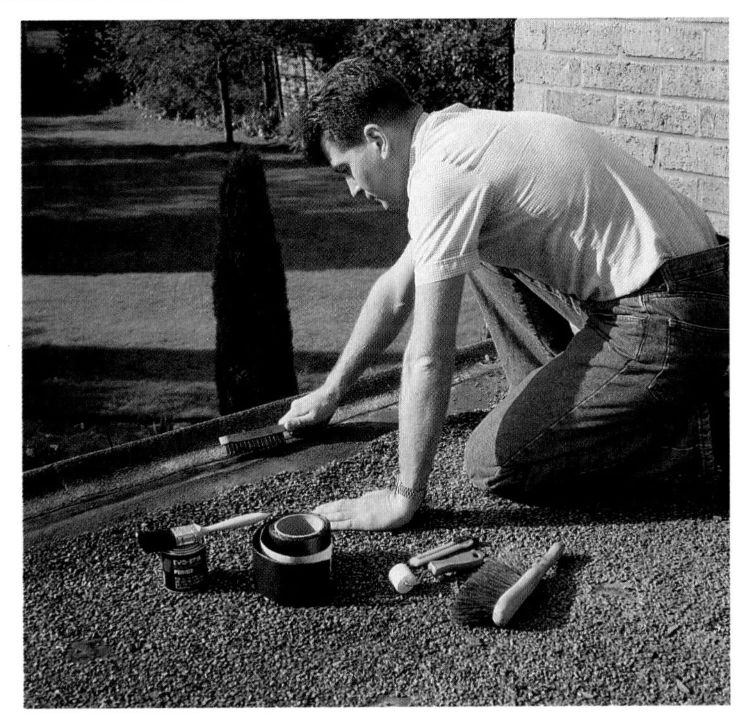

Fig 52 (*above*) Use self-adhesive flashing tape to repair leaky welted aprons. Brush back the chippings before applying it.

Fig 53 Stipple a second coat of waterproofer into the membrane, then apply a third when this is dry.

Fig 54 Finish off by pouring out a fresh layer of chippings and rake them out evenly.

Fig 55 If you are using a urethane-based liquid rubber sealer, apply the primer first with a wide paintbrush.

Fig 56 Brush on a thick coating of the sealer with a broom, taking it up over aprons and flashings too. It does not need chippings.

Repairing a Glass Roof

Glass roofs in lean-to buildings such as conservatories and covered ways are prone to damage from objects falling from higher up the building, and to accidental damage by carelessness from within. The glass in roofs of this sort should always be the wired type at the very least, and should ideally be toughened or laminated for extra safety. Otherwise there is a risk of serious injury if the glass is broken.

For this reason, it is sensible to replace any ordinary glass in a lean-to roof with safety glass, or to use clear plastic glazing materials as an alternative. The latter are easy for the do-it-yourselfer to cut to size (laminated safety glass has to be cut to size by a glass merchant, while toughened glass must be made to the size required) and offer much better resistance to shattering if something falls on to the roof.

Small-scale damage to the glass may simply result in a crack which will let in rainwater, while failed putty can allow water to penetrate along the lines of the glazing bars. Both can be repaired easily.

What to do

Where just a single pane is cracked, the best way of making a temporary repair is to use proprietary clear glass repair tape. Start by washing the glass to remove dirt and leave a clean surface to which the tape can adhere properly, then bed the tape along the crack. Work from the bottom upwards, overlapping successive lengths where the crack changes direction. Repeat the process on the underside of the roof.

If the pane is actually broken, it must be removed and replaced. If you can reach the pane from a step-ladder, chip away the putty and pull out the glazing sprigs to release it. Otherwise prop the roof from below before climbing on to it using crawl boards (*see* TIP). Slide the broken glass out, clean up the rebate and fit a new piece. Stop it sliding down by fixing a small strip of light metal to a roof timber if one is in line with the bottom of the pane, or, if not, use a strip-metal hook over the top edge of the next pane. Finally, waterproof the glazing bars with narrow strips of flashing tape.

What you need:

- clear glass repair tape
- sharp handyman's knife
- old chisel or glazier's hacking knife
- replacement pane
- putty
- putty knife
- glazing sprigs
- flexible strip metal
- gloves
- flashing tape

TIP

Always make sure that a glass roof is adequately supported from below by adjustable Acrow props and timber bearers before climbing on to it to carry out any repairs that cannot be reached from the eaves or vergé of the roof. Then lay crawl boards over the glazing bars immediately above the supports to carry your weight.

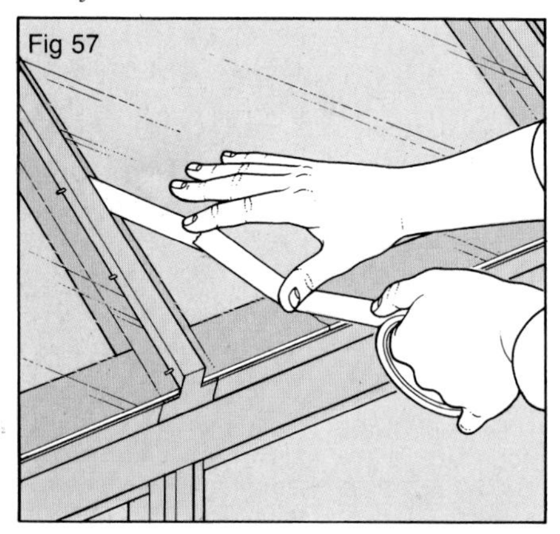

Fig 57 Repair minor cracks using clear-glass repair tape on both the upper and lower surface of the glass.

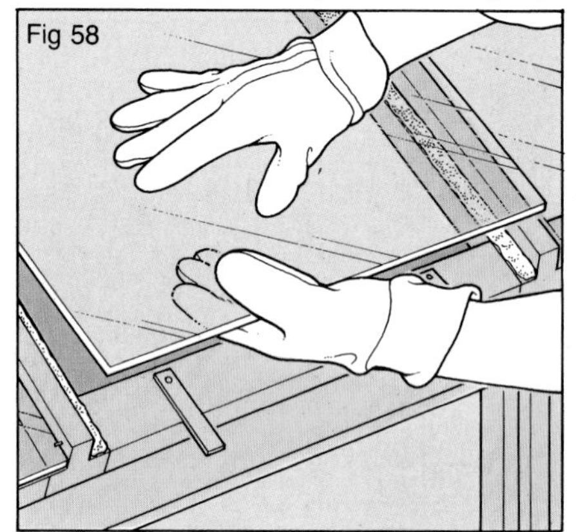

Fig 58 Remove a broken pane and slide in a replacement. Use a metal strip or hook to support its lower edge.

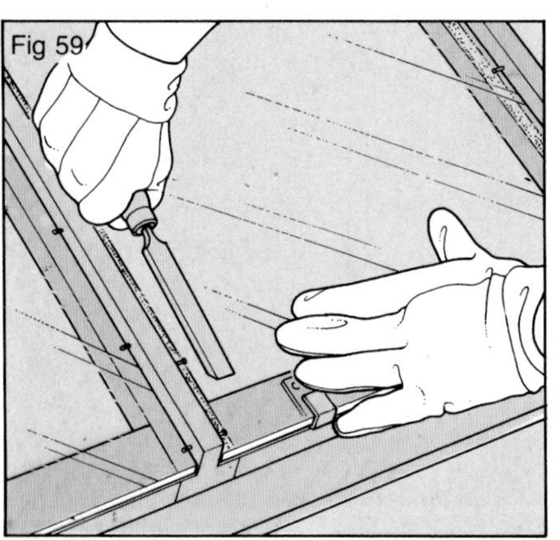

Fig 59 Secure the pane with glazing sprigs and replace the putty.

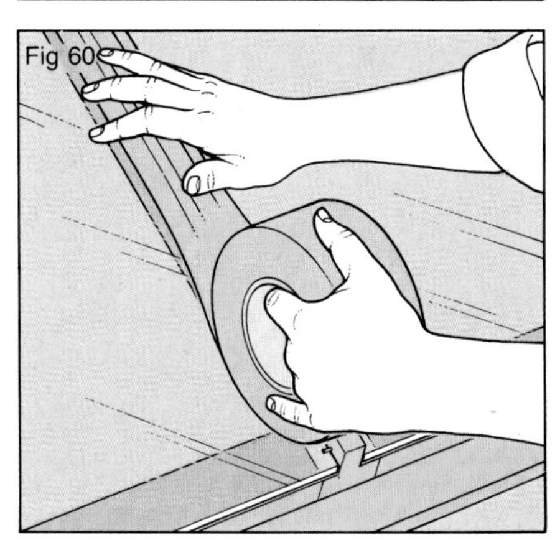

Fig 60 Waterproof the joints along the glazing bars by pressing on lengths of self-adhesive flashing tape.

Repairing a Corrugated Roof

Two sorts of corrugated-sheet roofing material – asbestos-cement and plastics of various types – are commonly used around the home, mainly on outbuildings such as garages and covered ways. They are laid with an overlap at the ends of the sheets, and with one or more corrugations overlapping along the edges, to resist water penetration and wind lift, and are secured to the supporting roof structure with hook bolts or screws (*see* CHECK). The nuts or screw heads should be protected by special sealing washers and caps. On ridged roofs made from asbestos-cement, the ridge is weatherproofed by a preformed ridge-strip. Where the roof abuts a wall, the top of the roof slope will be sealed with a flashing strip.

Both asbestos-cement and plastic sheets are prone to the same type of damage as glass roofs. In addition, asbestos-cement can weather and become dusty, with the risk of asbestos fibre being released into the air within the building, while some types of plastic sheet can discolour and become brittle with age.

What to do

If the sheet is actually holed, it should be replaced with a new sheet; repairs are rarely satisfactory except as a temporary measure. Gain access to the roof and use crawl boards (plus props below on lightweight structures) to support your weight, and release the fixings holding the damaged sheet in place. Slide it out in the direction of the eaves, and fit a replacement with the same size corrugations. Ensure that you have an adequate overlap at the top and sides and replace the fixings to complete the repair, dressing down flashings if these had to be lifted to release the old sheet. Dispose of old asbestos-cement sheet safely; contact your local authority's environmental health department for advice.

If old asbestos-cement sheet has become dusty and porous, use a high-pressure hose to clean the surface and then brush on two coats of heavy-duty bituminous waterproofer to seal the roof.

You can patch cracked plastic sheeting temporarily with clear-glass repair tape.

What you need:
- access equipment
- screwdriver or spanner
- replacement sheeting
- new fixings
- high-pressure spray equipment
- bituminous waterproofer
- glass repair tape

CHECK
- that you have replacement fixings of the correct type (*see* Fig 65)

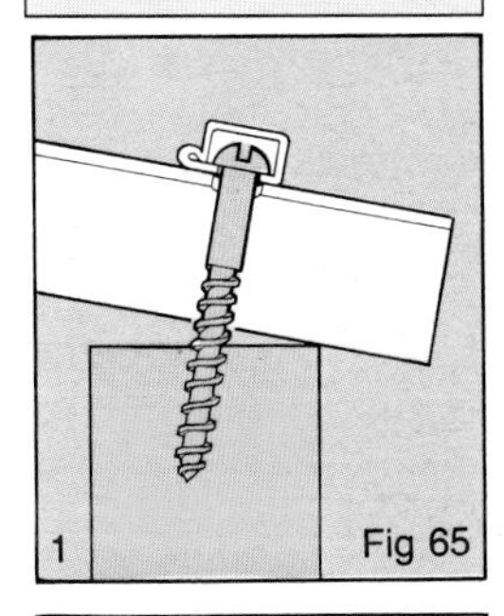

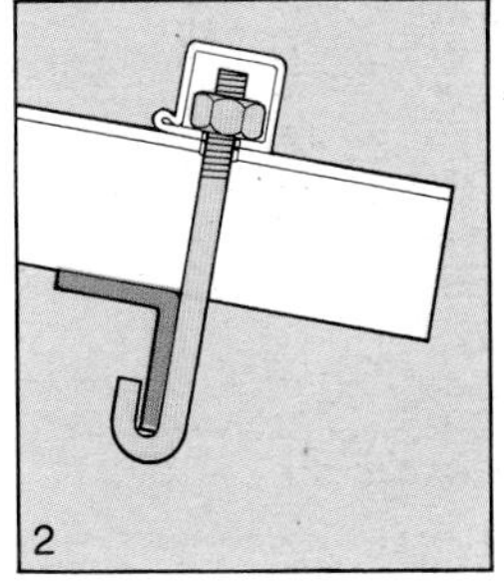

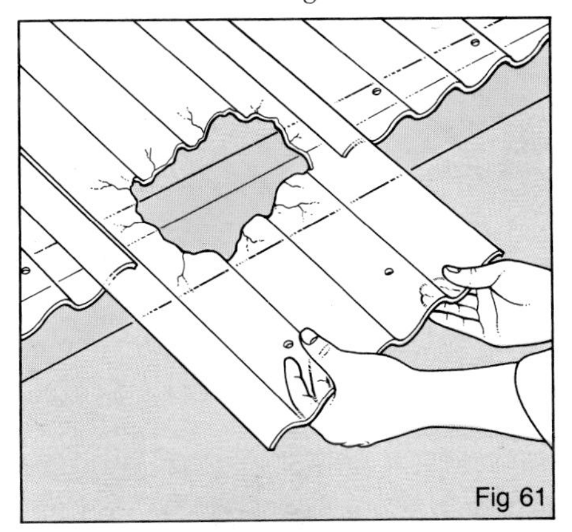

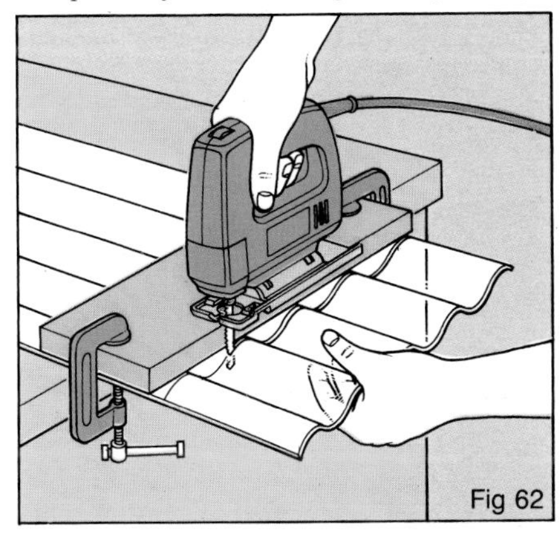

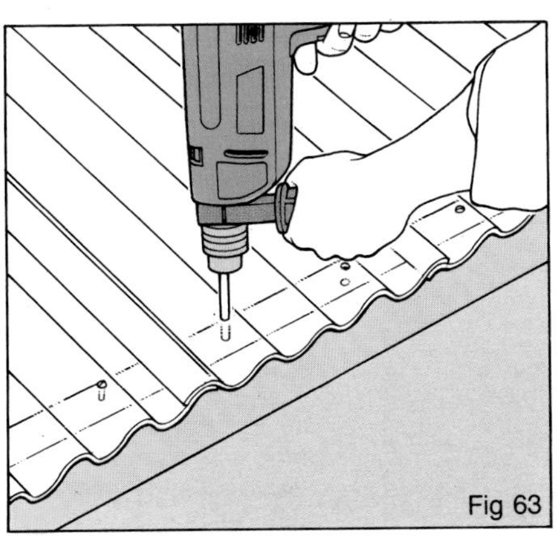

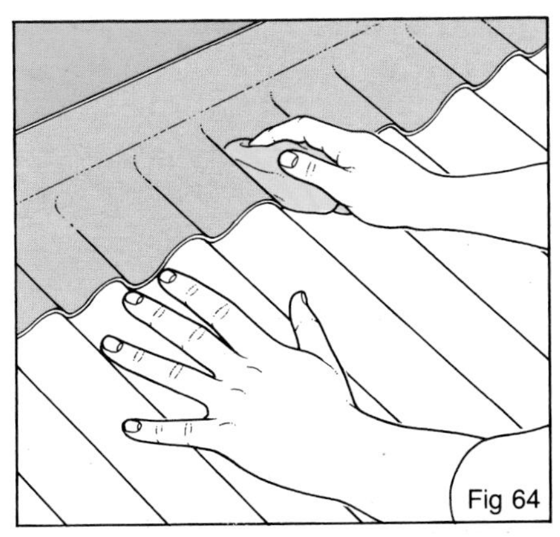

Fig 61 Remove the old fixings and release the damaged sheet.

Fig 62 Use a jigsaw and a guide batten to cut sheet.

Fig 63 Drill holes for the fixings and secure the new sheet.

Fig 64 Dress existing flashings neatly down over the new sheet.

Fig 65 Replacement fixings. **1** Screws for timber roofs. **2** Hook-bolts and nuts for metal roofs.

Repairing Flashings

Where flat or lean-to roofs meet a house wall, the junction between roof and wall is weatherproofed by a flashing strip. The best flashing material available is lead sheet, and this is still used on all top-quality building work. Its main advantage is that it can be easily shaped to conform to the profile of the roof surface, and it is extremely durable, although it can be lifted and torn by high winds. Felt flashings are less durable, being prone to cracking and splitting, but are often used as an economical alternative to lead on flat roofs. Mortar flashings are the least satisfactory of all materials because they soon crack due to differential movement between wall and roof, and can quickly become porous.

When flashings fail, water can penetrate at the junction between roof and wall and may then run down roof timbers to emerge at a point indoors remote from the source of the leak. For this reason, it is worth inspecting the adjacent roof surfaces for faults at the same time that the flashing repairs are carried out.

What to do

The commonest problem with **lead flashings** is lifting due to high winds. These may pull the upper edge of the flashing out of its chase in the brickwork, and may also tear the metal.

Where the flashing has been torn from its chase but is otherwise undamaged, it can be replaced relatively easily. Start by lifting the whole length of flashing away. Then rake out the old mortar from the length of the chase, saving any wedges of lead used to hold it in place, and offer the flashing back into position. Tuck the lead wedges into the chase to hold the top of the flashing in place. If you have none, use slips of slate or similar material instead. Then mix up a quantity of 1:5 cement:sand mortar (or buy a small bag of dry ready-mixed repointing mortar), and fill the chase, pointing up the finish neatly. Then dress the replaced flashing neatly down over the roof surface using a block of wood and a hammer.

If the flashing is split but is otherwise

What you need:

- cold chisel
- club hammer
- pointing trowel
- 1:5 repair mortar
- bituminous mastic
- filling knife
- self-adhesive flashing tape
- flashing primer
- sharp handyman's knife
- wallpaper-seam roller

CHECK

- that lead flashings are well dressed down on to roof surfaces to prevent high winds from lifting them
- that self-adhesive flashing tape is firmly bonded to the wall surface, with no air bubbles trapped beneath it

TIP

In cold weather, store flashing tape indoors for an hour before using it, to make it more flexible and easier to apply.

Fig 66 Where high winds have pulled lead flashing away from the wall, lift the length off and rake out the chase in which it was bedded. Save any lead wedges you find.

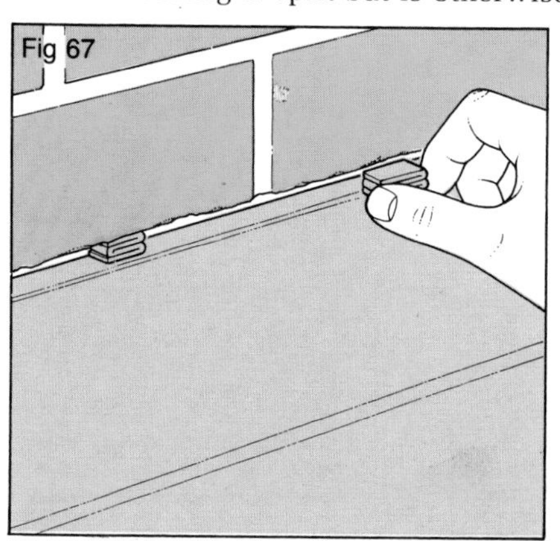

Fig 67 Reposition the flashing and push in the folded wedges to hold it in place.

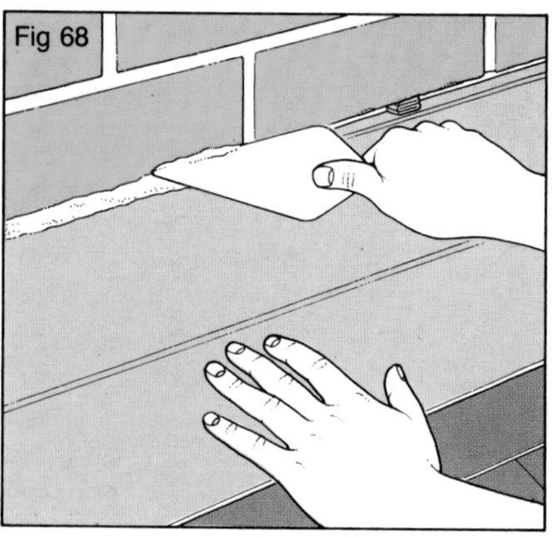

Fig 68 Use 1:5 mortar to repoint the chase.

Fig 69 Repair any small splits or tears in lead flashing with bituminous mastic.

intact, you can use a generous layer of bituminous mastic to make it waterproof. If the surface of the flashing simply appears pock-marked and weathered, give it a coat of bituminous waterproofer.

With **felt flashings** that are torn or damaged, it is best to cut them away and to replace them completely with a new flashing using self-adhesive flashing tape. After removing the old flashing, brush a coat of the special flashing primer (sold with the tape) on to the wall and roof surface along the line of the flashing, and leave it to become tacky. Then cut the flashing tape to length, peel back one end of the backing paper and press the end of the tape into place. Peel off more of the paper and bed the tape along its length, pressing it firmly against both the wall and roof surfaces to ensure a watertight bond. Run a wallpaper-seam roller along the top edge of the flashing to apply extra pressure at this key point.

Hack away all old **mortar flashings** completely, and replace them with flashing tape after treating the wall and roof surfaces with flashing primer.

Fig 70 (*above*) Flashing tape is ideal for repairing flashings in a wide range of situations.

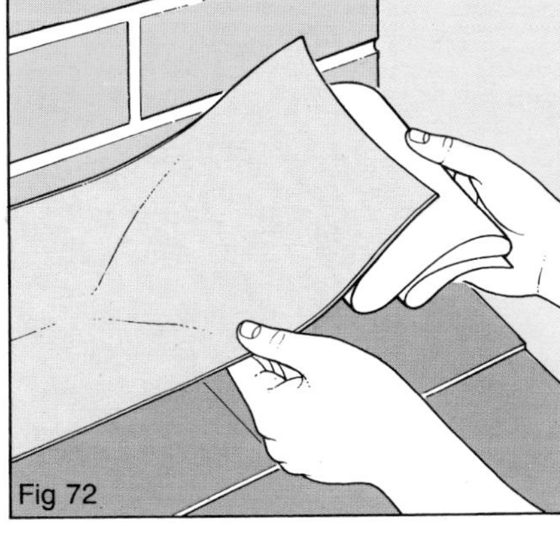

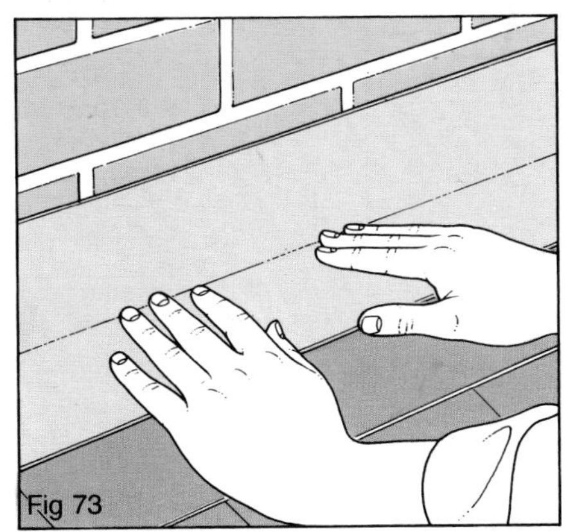

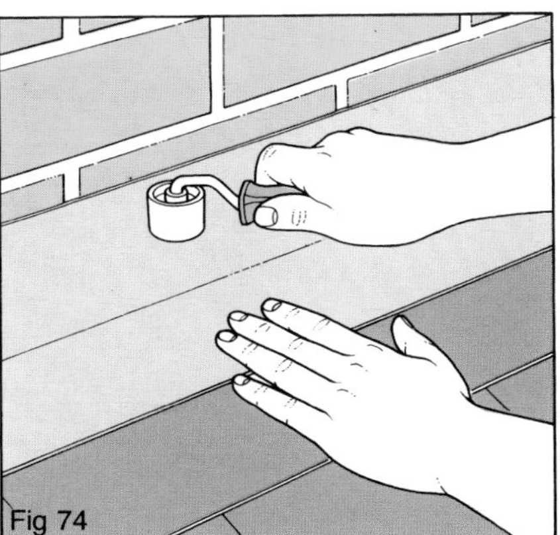

Fig 71 Remove old felt or mortar flashings and brush flashing primer on to the wall and roof surface. Leave it to become tacky.

Fig 72 Peel off the backing paper and press the tape into place.

Fig 73 Apply even pressure to the tape to bond it firmly to the roof surface below.

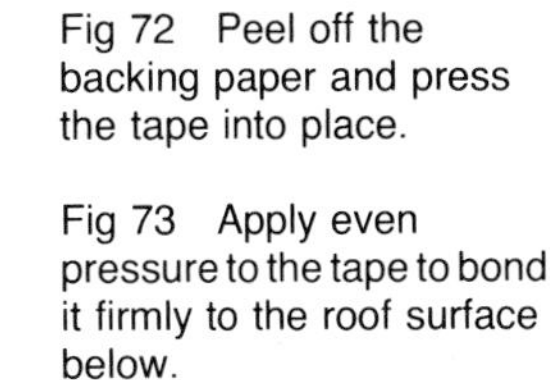

Fig 74 Use a wallpaper-seam roller to ensure waterproof seal between tape and wall.

Repairing Chimney-Stacks

Chimney-stacks are a potential weak spot in the weatherproof exterior of the house, for two reasons. Firstly, the stack itself can suffer from defective brickwork, failed pointing or an insecure pot, all of which can allow water to penetrate the structure and give rise to damp patches on chimney-breasts and ceilings inside the house. Furthermore, because the stack breaks through the roof slope, the junction between the roof and the stack must be kept waterproof, and the complex flashings needed to do this can deteriorate with age or may be damaged by high winds.

The first step in assessing what needs to be done is to make a visual inspection of the stack. You can do this from ground level with binoculars, but it is better to examine it at least from eaves-level by putting up your ladder with a ladder-stay as close to the stack as possible. If you are happy at heights and the design of the roof allows it, you can put up a roof ladder alongside the stack and carry out your inspection at close quarters. If you are not, leave it to an expert.

What to do

If you are carrying out the work yourself, it is safest to set up a roof ladder as described on page 24 and then to use the components of a platform tower to build a working platform around the stack. You will then have a base on which to store tools and materials, plus the added safeguard of a handrail and toe boards all round. Never emulate professional roofers and try to climb directly on the roof slope; one slip could kill.

Start the repair work at the top. If the pot is loose and the sloping mortar (called flaunching) securing it to the stack is cracked or missing, secure the pot to the stack with rope. Then chop away the old flaunching with a brick bolster and rebed the pot in fresh mortar, using a 1:5 mix with added plasticizer. Use pieces of slate or fibre-cement sheet to support the pot if necessary, and slope the flaunching down from the pot to the edge of the stack to assist rainwater run-off. If the flaunching is sound but has one or two minor cracks

What you need:

- access equipment
- rope
- brick bolster
- club hammer
- pointing trowel
- 1:5 repair mortar
- exterior mastic
- silicone water-repellent sealer

CHECK

- that all access equipment is securely fixed and roped to the building
- that you rope the pot to the stack before removing the flaunching

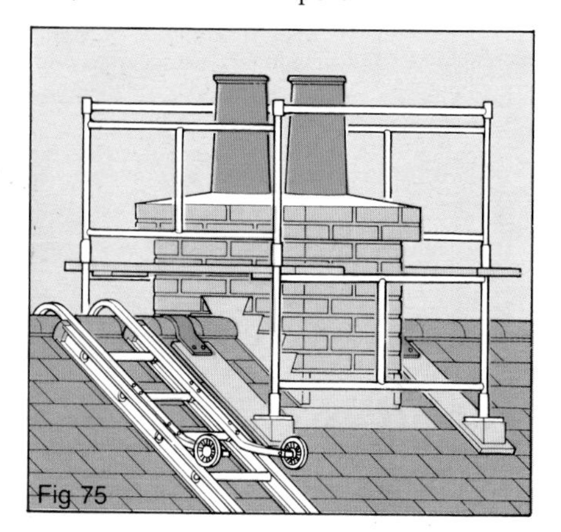

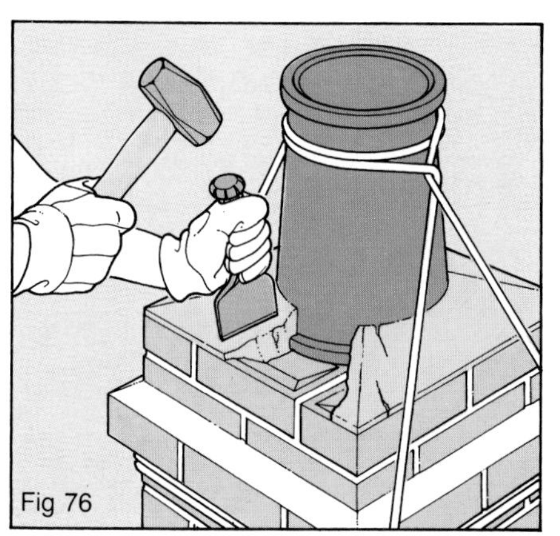

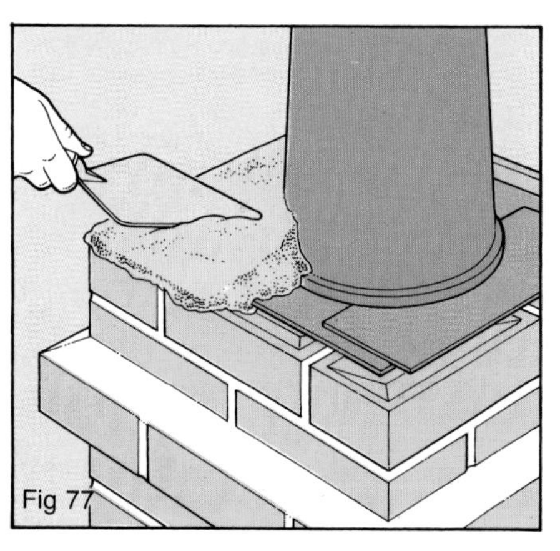

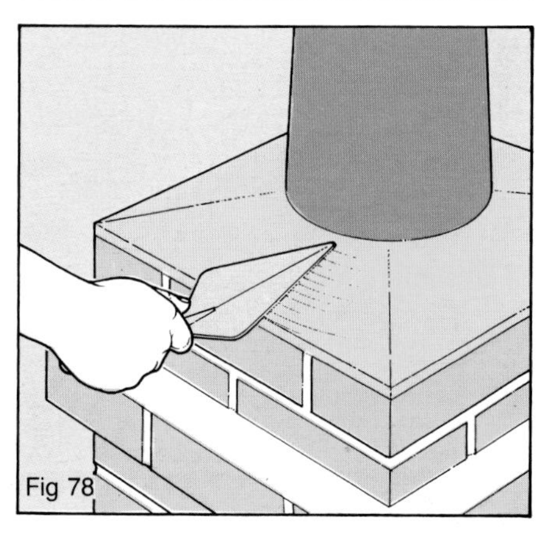

Fig 75 Before carrying out any work on the stack, it is safest to erect a solid working platform round it using slot-together platform tower components. Fit toe boards and a guard-rail all round.

Fig 76 Rope the pot to the stack so it cannot move, and chop away defective flaunching with a brick bolster and club hammer.

Fig 77 Support the pot on slates if necessary, then remake the flaunching with 1:5 repair mortar.

Fig 78 Finish the flaunching smoothly with a slope to encourage rainwater to run off.

in it, fill them with exterior-quality mastic as an alternative to using mortar.

Next, check the condition of the pointing on the stack. Rake out any loose material and repoint with 1:5 mortar, finishing off the joints with weathered profiles to minimize rain penetration. Pay particular attention to the mortar fillet above any projecting brickwork (called *corbelling*) on the faces of the stack, to ensure that the slope will help throw rainwater clear of the stack.

You may find that the stepped flashings down the sides of the stack, and those across its back and front, are loose or torn. Rake out the pointing, wedge the tongues back into their chases and repoint (see page 34 for more details). If the flashing is torn or split, either repair it with bituminous mastic or stick on a patch of self-adhesive flashing tape. Check that the lower edges of the flashings are dressed down on to the roof surface, using a block of wood and a hammer to shape them. Finally, give the stack brickwork a coat of silicone water-repellent sealer.

Fig 79 (*above*) Sound pointing and smooth flaunching are the hallmarks of a well-maintained stack.

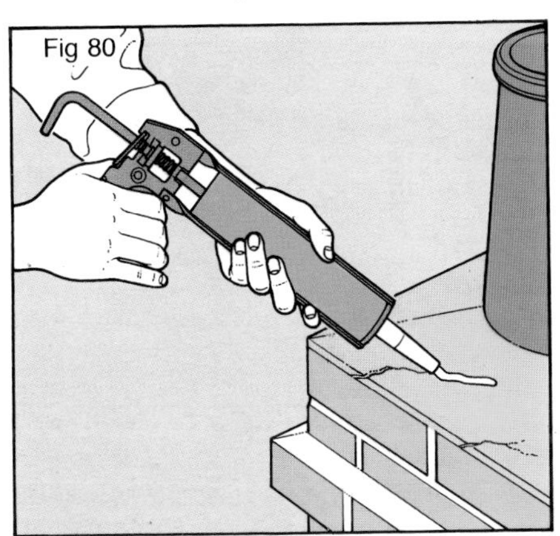

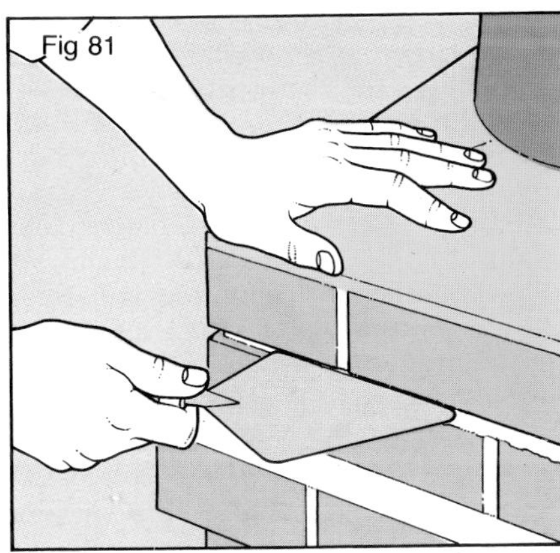

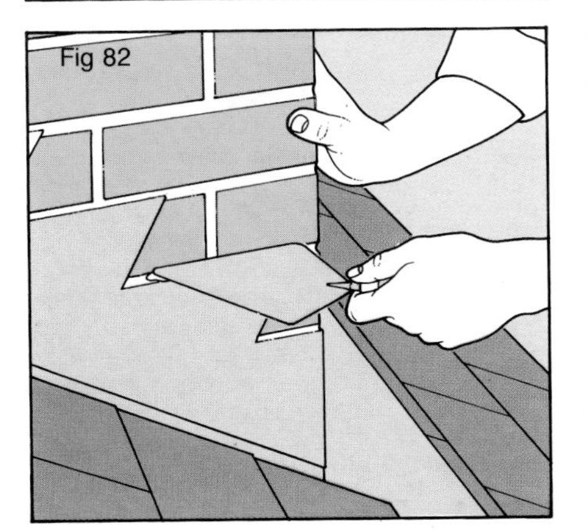

Fig 80 If there are just one or two minor cracks in the flaunching, fill them with exterior-quality mastic.

Fig 81 Rake out any defective pointing on the stack, and repoint with 1:5 repair mortar.

Fig 82 Check that flashings around the stack are secure in their chases, and repoint if necessary.

Fig 83 Waterproof the stack with a coat of silicone water-repellent sealer.

Dealing with Disused Flues

Many older homes in this country were built with a fireplace in all the main habitable rooms in the house, and with a corresponding number of flues and chimney-stacks to serve them. It is certainly true to say that open fires are now more popular than in the 1960s and 1970s, and most new homes now have a fully functioning fireplace in at least the main living room, but few people want to light one anywhere else. This means there are likely to be several unused flues, and possibly their fireplace openings too, which would be better blocked off to cut down on unnecessary heat losses and uncomfortable draughts, and to reduce water penetration down the open stack.

It is vital to remember that if the flue is being left intact it must not be sealed off completely or condensation will occur within it. This can then leach through the stack brickwork, carrying with it tar from the flue-lining, which can stain decorations and damage plasterwork. The solution is to provide ventilation openings at both the top and bottom of the flue.

Fig 84 (*above*) Clean up removed ornamental pots for use as garden ornaments.

What to do

Start by tackling the top of the chimney-stack. You have a number of options here, the simplest being to remove the chimney-pot from the top of the disused flue and to bed a half-round ridge-tile on a mortar bed over the flue opening. This will keep rain out while still allowing good ventilation, but you may need to experiment with the orientation of the tile to the prevailing winds to minimize wind noise. This option is ideal where just one flue of a multi-flue stack is disused.

Where the entire stack is disused, it is better to cap it off completely. Start by removing the pots and flaunching. Then chop out a brick from two opposing faces of the stack and insert air-bricks in their place to ensure good ventilation. Finally, bed one or more paving slabs in mortar on top of the stack to seal it.

If the stack brickwork is in poor condition or shows signs of leaning owing to sulphate attack, it is best to demolish it down to near roof-level before capping it. Set up a safe working platform around the stack as described on page 24, and lower the waste material carefully to the ground in buckets. Remember to include air-bricks near the top of the lowered stack.

What you need:

- club hammer and brick bolster
- ridge-tile or paving slab(s)
- mortar
- pointing trowel
- air-bricks
- access equipment

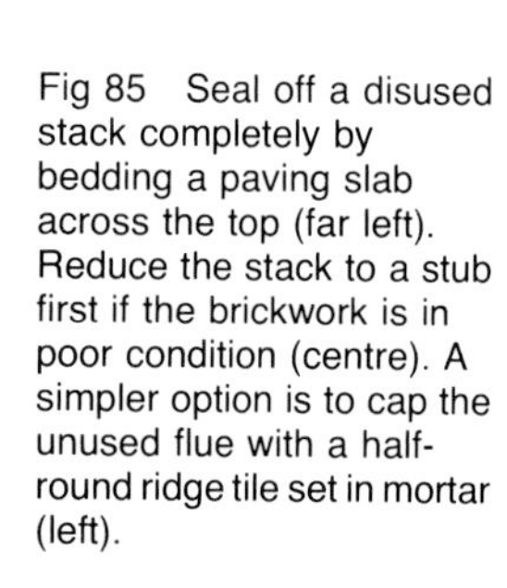

Fig 85 Seal off a disused stack completely by bedding a paving slab across the top (far left). Reduce the stack to a stub first if the brickwork is in poor condition (centre). A simpler option is to cap the unused flue with a half-round ridge tile set in mortar (left).

The Fireplace Opening

Once the top of the flue has been dealt with, you can turn your attention to the disused fireplace opening indoors. Start by removing the fire surround and raised hearth if these are still in place. The hearth will probably be a slab of reinforced concrete cast on top of the so-called constructional hearth that finishes at floor level, and it is generally fairly easy to prise this up with a crowbar or a garden spade. Get help in carrying it out of doors as it will be very heavy.

Next, turn your attention to the surround itself. This will probably be secured to the wall by lugs buried in the plaster at each side, so chop some plaster away to expose them. Lever out the fixings and lift the surround away, in one piece if necessary, or by removing first the mantelpiece and then the sides if it is clearly in sections. Do not demolish it if it is attractive or collectable; offer it to a local architectural salvage merchant or advertise it for sale so someone else can use it for a restoration project.

Unless the fireback is in poor condition, it is better to leave it in position than to remove it; then the opening can be easily brought back into use in the future if desired. Simply brush out the opening and block it off, either with plasterboard on a timber subframe or, for a more permanent job, with bricks or blockwork.

If you decide to use plasterboard, nail a frame of sawn timber to the sides of the opening, set back by the thickness of the board, and then fix the sheet in place before giving it a skim coat of plaster. Do not forget to cut a ventilation opening near ground level; this can then be covered with a plastic ventilator grille.

If you are using masonry to block the opening, build it up course by course. Again, remember to include an air-brick low down in the infill panel to ensure a free flow of air through the flue. Then plaster over it to leave the finished surface flush with the surrounding plasterwork, and fit a plastic grille over the air-brick to complete the job.

What you need:
- demolition tools – crowbar, club hammer, brick bolster, spade
- sawn timber, masonry nails, plasterboard, plasterboard nails *or*
- bricks or blocks, mortar, plaster
- air-brick
- ventilator grille
- sundry tools

CHECK
- that there is an up-draught through the flue by holding a lighted taper near the low-level ventilation opening

TIP
Make sure that the top of the stack is capped before sealing the fireplace opening, in case a bird falls down the open flue.

Fig 86 To block the fireplace opening with plasterboard, use masonry nails to fix a sawn timber sub-frame around the inside of the opening.

Fig 87 Then nail on the plasterboard, ready to be plastered over. Remember to cut a ventilation opening in the board, and cover it later with a grille.

Fig 88 If you are using masonry, build it up course by course, incorporating an air-brick low down.

Fig 89 Finish the job by plastering over the infill, and fit a grille over the air-brick.

Repairing Eaves Woodwork

The eaves on the vast majority of houses, whether they have pitched or flat roofs, are finished off with fascia boards to enclose and protect the ends of the roof rafters. If the eaves project beyond the face of the walls below, the undersides of the projections are filled in with horizontal boards called soffits. At gable ends, the roof is finished off with two barge-boards in the shape of a letter A, and again soffits are used where the gable end of the roof projects.

In most cases, all these boards are of softwood, although other materials such as asbestos-cement sheet have been used with varying degrees of success in the past, and some modern homes have eaves fitted out with rigid cellular plastic boards.

The big drawback with softwood is that unless it is thoroughly treated with wood preservative and kept well protected from the elements by paint or another finish, rot soon sets in. This is not only ugly; it can also allow rot to spread to the roof timbers, and may lead to the collapse of guttering attached to the fascia.

What to do

If rot is present only on a small scale, it may be possible to treat the affected areas with a rot repair kit. This consists of a liquid resin, which is brushed on to the rotten wood to harden it, and a special filler to repair missing areas. However, this method will be expensive for more than local repairs and, if the attack is extensive, it is quicker, simpler and cheaper to replace the affected timbers with new wood.

Start at one end of the eaves, using a crowbar or a claw hammer to prise away the timbers. It is usually easier to remove the soffits first, unless they are tongue-jointed into grooves in the rear edge of the fascias, in which case tackle the fascias first, after taking down the guttering. When you have prised off all the rotten boards, pull out any old fixing nails remaining in the joist ends.

Always replace eaves timbers with wood that has been pre-treated with preservative by pressure or vacuum impregnation, so

What you need:

- crowbar or claw-hammer
- pencil and scribing block
- jigsaw
- spirit-level
- screwdriver
- replacement timber
- wood preservative
- galvanized nails
- access equipment

CHECK

- that all wood has been pre-treated with wood preservative
- that cut ends are given a brush coat of preservative for extra protection
- that your loft has good cross-ventilation. *See* page 59 for details on how to install eaves ventilators

TIP

When you have removed the old fascia and soffit boards, treat the exposed ends of the roof joists with wood preservative to guard against future rot attack.

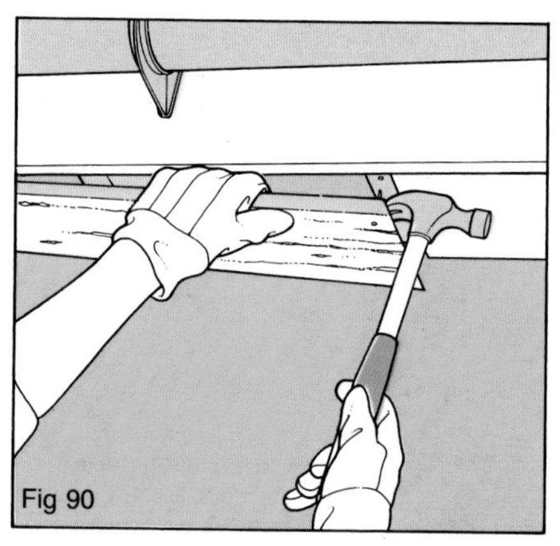

Fig 90 Start by prising off all the rotten boards with a claw-hammer or crowbar.

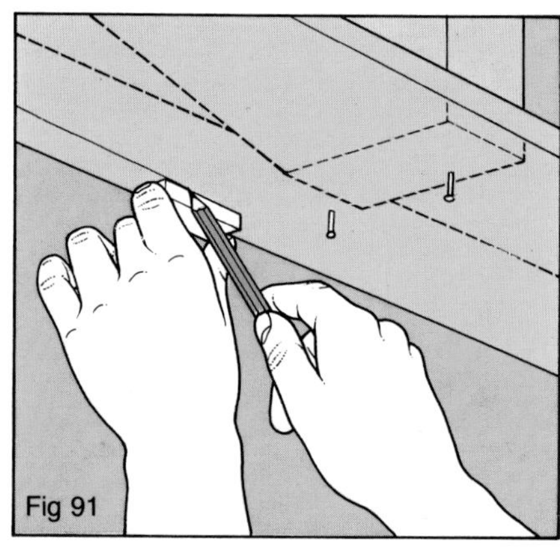

Fig 91 Pin each new soffit board temporarily in place, and scribe the wall's profile on its rear edge with a block and pencil.

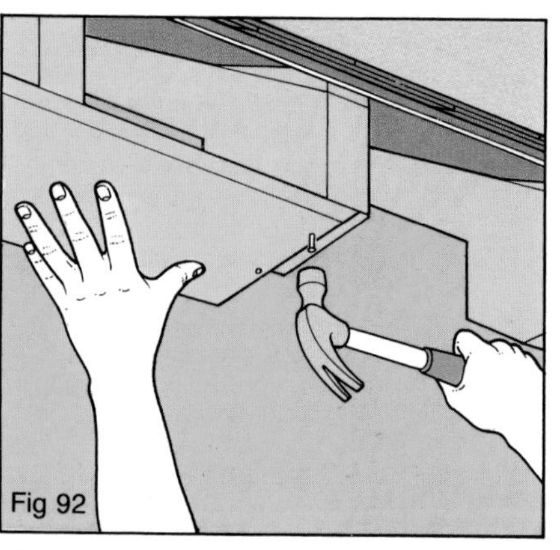

Fig 92 Cut the board ends at 45° for a neat joint, and nail each board to the rafter ends.

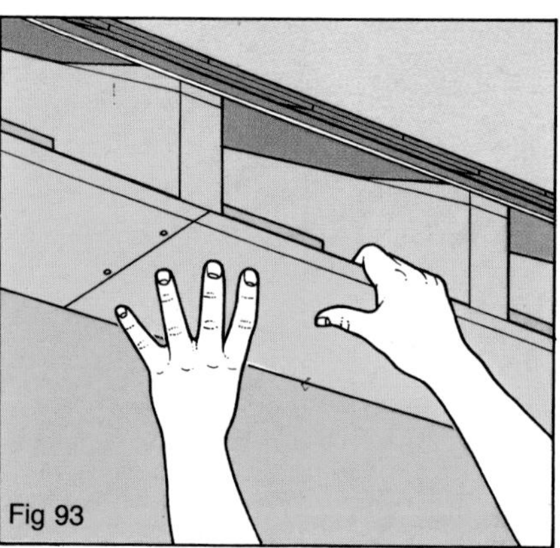

Fig 93 Butt the next length of board up against its neighbour, and fix it in the same way.

that it stands a reasonable chance of repelling future rot attacks.

Start replacing the timber by pinning an over-wide length of wood to the underside of the eaves, and use a scrap of wood as wide as the overlap beyond the rafter ends to scribe the profile of the wall on to the back edge of the board. Cut along the scribed line and offer the board up in position, nailing it to each rafter end.

Cut the board ends to 45° using a jigsaw to create a sloped joint, and fix successive boards in the same way after first scribing them to width. Brush preservative on all cut ends before fixing the boards.

Next, offer up the lengths of fascia board, checking that they are level and are tucked behind the underfelt, and nail each one to the rafter ends. Use galvanized nails so they will not rust.

If you are also replacing barge-boards, use the old boards as templates for cutting the new ones to size. Screw or nail them in place, and finish off by attaching triangular tailpieces to the bottom ends of the boards to close off the gap at the end of the soffit board.

Fig 94 (*above*) Treat new fascias with micro-porous paint for minimal future maintenance, before replacing the guttering.

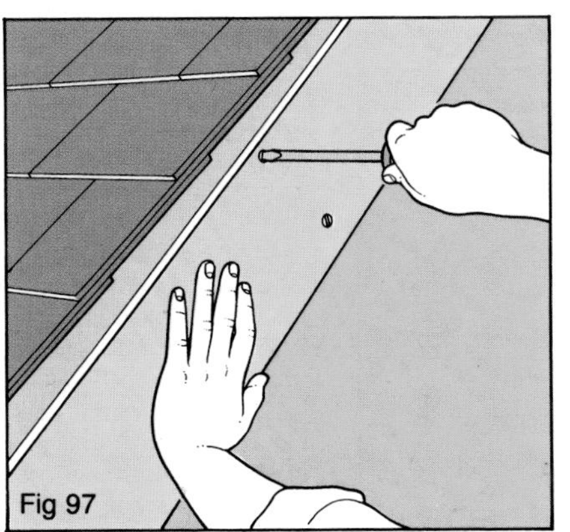

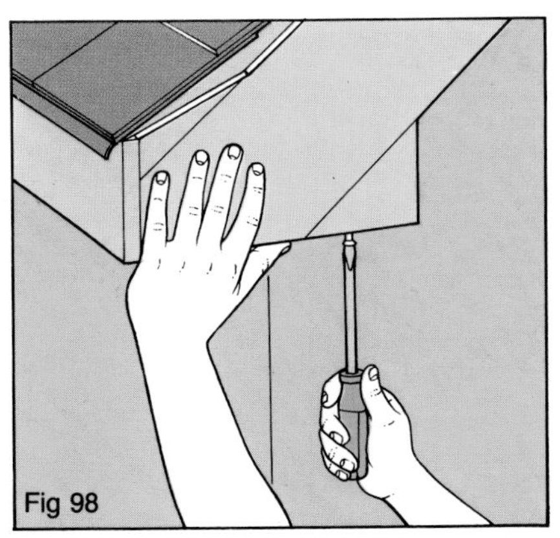

Fig 95 Offer up the first length of fascia board, checking that it is level and is tucked behind the underfelt.

Fig 96 Nail the board to each rafter end in turn, using galvanized nails.

Fig 97 Screw or nail new barge-boards to gable ends, butting them tightly up against the undercloak.

Fig 98 Complete the job by fitting triangular tailpieces to the bottom edge of the barge-board to cover the end of the soffit.

Repairing Gutters

Gutters suffer from a variety of common faults. Rust is the greatest enemy of old cast-iron guttering, but leaking joints and sagging brackets can cause trouble too. Physical damage from carelessly wielded ladders is the most frequent problem with modern plastic types.

It is important to put gutter problems right promptly, since faults can allow water to run down the house walls and cause penetrating damp (not to mention unsightly staining). Therefore, if you suspect a problem, set up your ladder – with a ladder-stay if you have projecting eaves – so that you can climb up and inspect things at close quarters.

You may find that an overflow is caused by nothing more serious than a build-up of silt and debris in the gutter, especially at corners. It is a simple matter to scoop these blockages out with a garden trowel or a gloved hand, and then to rinse the system through from its highest point with clean water from a garden hose. However, any physical defects you find will need a little more effort to put right.

What to do

If you have cast-iron gutters which have rusted through, the only solution is to replace the affected sections. For just one or two lengths, it is best to replace like with like, but if more than half the system is beyond repair it makes more sense to remove everything and to install a new plastic system from scratch.

To replace a length, you will have to undo the fixing-nut and bolt at each end. Since these will have rusted solid years ago, free them by hacksawing through the nut beneath the joint. If the gutters are the flat-backed ogee type fixed directly to the fascia, try freeing the fixing screws with a blow from a hammer. Should that fail, it is best simply to drill off the screw heads. You will need a helper on another ladder to lift the length down, as it will be too heavy and unmanageable for one person to handle safely.

Clean up the joints on adjacent sound lengths of gutter with a wire brush, and line the socket with mastic to make a

What you need:

- hacksaw
- screwdriver
- hammer
- wire-brush
- mastic
- filling knife
- adjustable spanner
- gutter repair bracket
- bituminous paint
- self-adhesive flashing tape or
- glass fibre repair kit

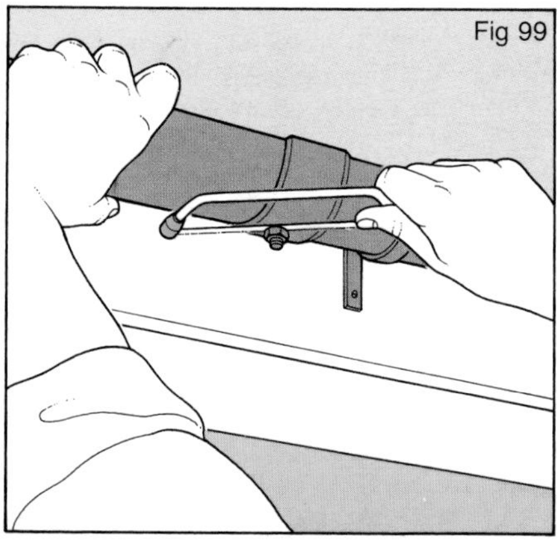

Fig 99

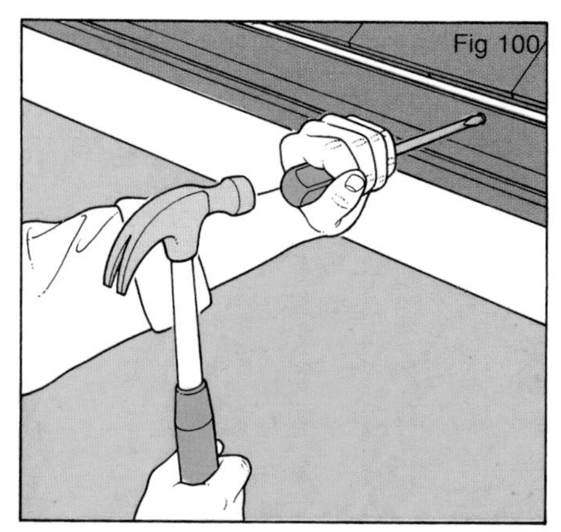

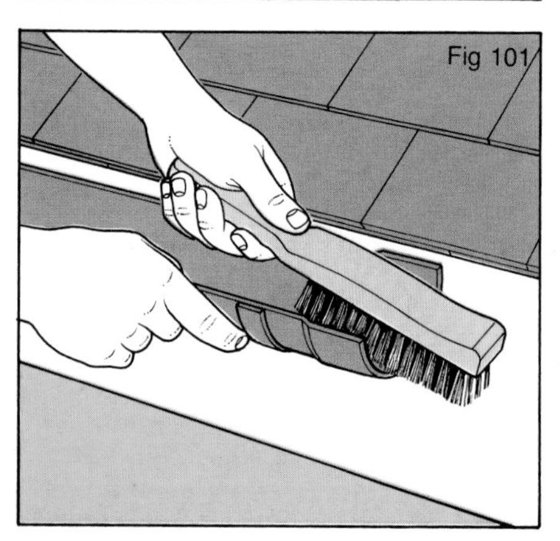

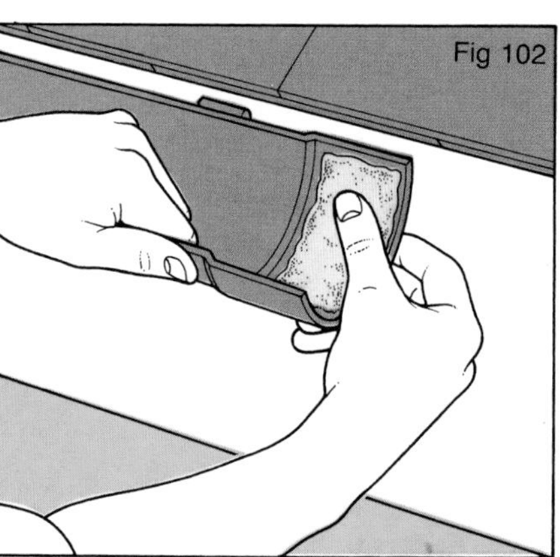

Fig 99 Hacksaw through the fixing nut to free the joint between lengths of cast-iron guttering.

Fig 100 If screws securing ogee guttering will not undo, give the screw head a sharp blow with a hammer to free it.

Fig 101 Once you have removed the damaged section, clean up the ends of adjacent lengths with a wire-brush.

Fig 102 Bed some mastic in the socket end of the guttering to make a watertight joint.

Fig 103 Remake the joint with a machine screw; you will not be able to operate a spanner inside the gutter (*right*).

Fig 104 Remove any excess mastic from the joint to leave a smooth repair.

watertight joint. Offer up the new section and fit a new machine screw and nut. Then scrape off any excess mastic that has oozed out to leave a neat finish. With ogee types, fix the new length to the fascia board first, then remake the joints at each end of the length. If the length is sagging because a bracket has failed or pulled out of the fascia, you can make a temporary repair by driving in a large wood nail below the sagging section. Repair the sag permanently with a special repair bracket which is screwed to the fascia from below the gutter.

Once you have completed these repairs, give the inside surface of the gutter a couple of coats of black bituminous paint to give it some protection against future rust attacks.

If you have old asbestos-cement gutters that are porous and in poor condition, it is best to replace them completely. Wear a face-mask, goggles and gloves as you dismantle the old system, and contact you local authority's environmental health department for advice on disposing of the debris safely.

Fig 105 (*above*) Simply replace a badly damaged length of plastic gutter by snapping a new section into the connector.

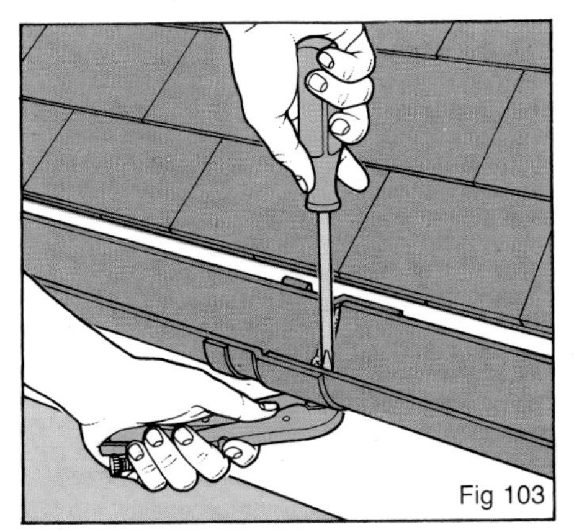

Fig 103

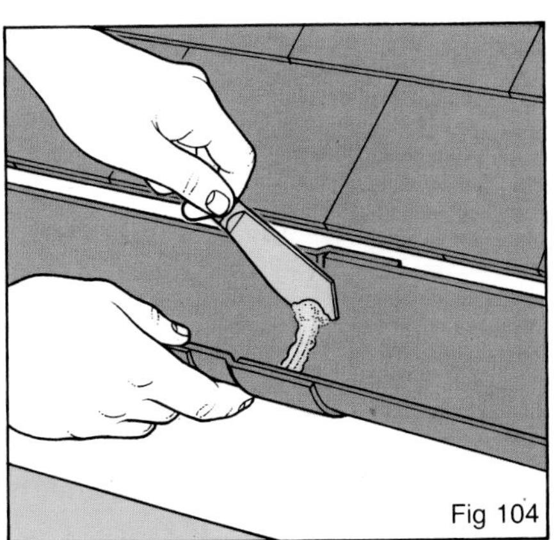

Fig 104

Repairing Cracked Gutters

Cracks and splits in all types of guttering can be repaired with self-adhesive flashing tape or, for a more permanent repair, with a glass-fibre car-repair kit. With flashing tape, clean the inside of the gutter and apply flashing primer to the damaged area. Then cut a piece of tape to size and simply press it firmly into place.

With a repair kit, wire-brush the rusted area back to bare metal, cut two patches of glass-fibre mat and bed them in the resin over the crack, one on top of the other, following the repair kit manufacturer's instructions carefully.

> CHECK
> • that lengths of cast-iron gutter are adequately supported before freeing the connections between lengths, as these may be all that is holding the length up. Have a helper alongside you on another ladder for safety

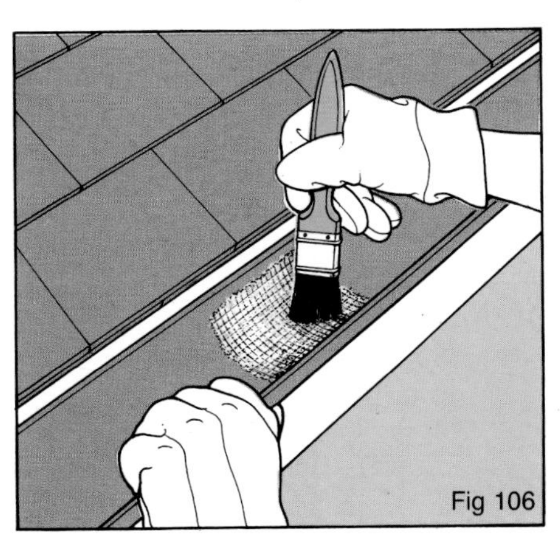

Fig 106 Repair cracks or holes in all types of gutter with a piece of self-adhesive flashing tape, or use mesh and resin from a glass-fibre car repair kit.

Repairing Downpipes and Hoppers

If your house suffers from problems with the guttering, it is likely to have similar faults with the rest of the external plumbing, i.e. the downpipes that carry rain-water away to the underground surface-water drains, and in older properties with two-pipe waste systems, the hoppers and waste pipes which collect water from upstairs appliances and channel it via separate downpipes to the sewers.

As with old gutters, the commonest problems with downpipes are blockages and leaks. Blockages occur because debris is washed into the downpipes from the gutters – a good reason for cleaning your gutters regularly – and these can cause water to overflow from joints in the pipe. Leaks can also occur if the downpipe rusts through or is physically damaged (the latter fault is more common with plastic than cast-iron pipe). Hoppers suffer mainly from blockages, caused by a combination of wind-blown debris and a build-up of soapy residues inside, they can be quite evil-smelling and so are no longer used on modern waste systems.

What to do

If you suspect a blockage in a downpipe, you may be able to clear it using your garden hosepipe as a combination of drain rod and pressure jet (*see* CHECK, right). Set up a ladder to the head of the pipe, lift off the double elbow (called the swan's neck), which connects the gutter to the downpipe, and insert the hose. Push it down the pipe until it meets the blockage, then turn the water on full blast.

If this fails to shift it, you have little option but to dismantle the pipe. Start at the bottom, and use a claw-hammer or crowbar to lever the pipe nails away from the wall. Lift each section away, taking care not to drop upper sections as you free them; it is best to have a helper on another ladder alongside to take the weight as you remove the fixings. Then rod through the blocked length and reassemble the run, using screws and wall-plugs rather than more nails to secure the brackets to the wall. Finally replace the swan's neck at the top.

What you need:

- access equipment
- garden hose
- claw-hammer or crowbar
- drain rods
- screws and wall-plugs
- drill and masonry bit
- screwdriver
- self-adhesive flashing tape and primer
- aerosol paint
- new pipe and fittings
- pipe adaptor

CHECK

- that your garden hose is fitted with a check-valve at the tap connection. This is required by the new water bye-laws to guard against the risk of foul water being sucked back into the mains supply if there is a drop in mains water pressure

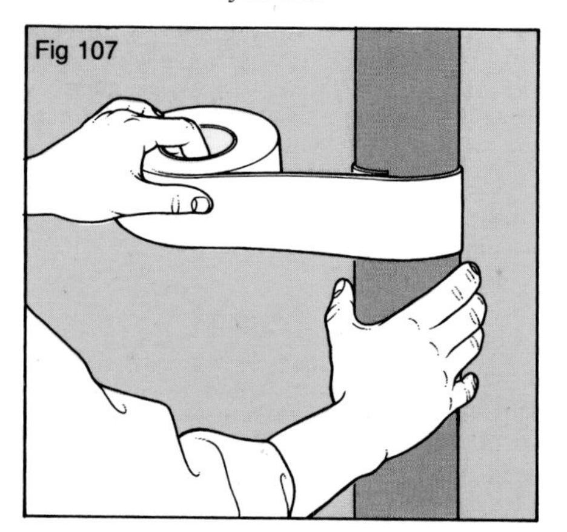

Fig 107 Repair localized holes or cracks in the pipe temporarily by wrapping self-adhesive flashing tape around it.

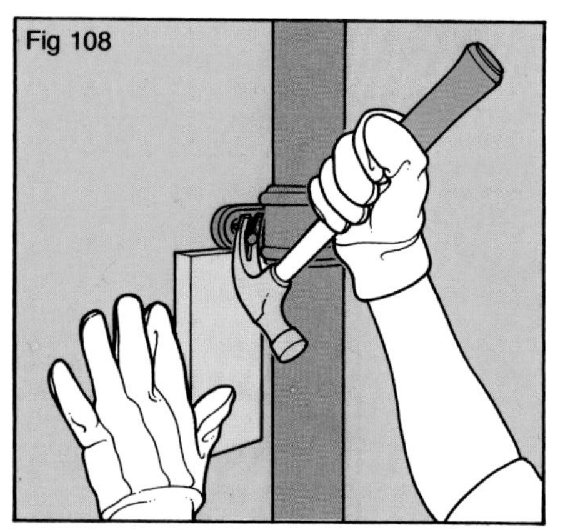

Fig 108 If you cannot clear a downpipe blockage with a hose, dismantle the pipe run from the bottom upwards. Use a hammer or crowbar to prise out the pipe nails.

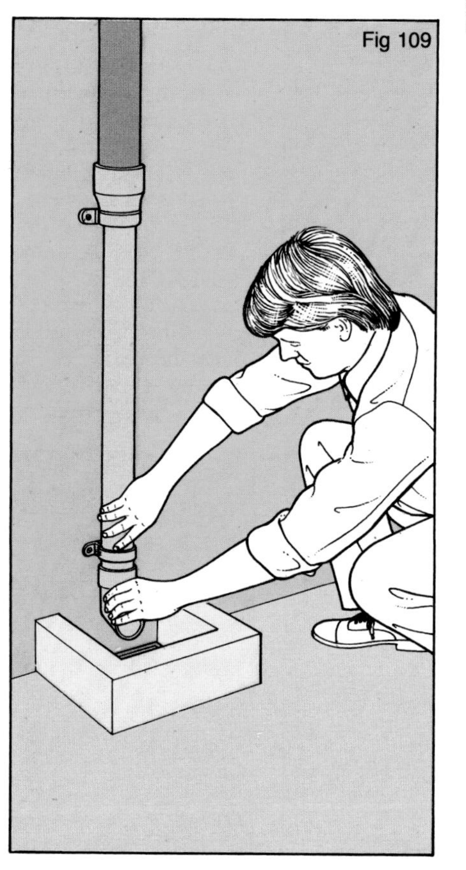

Fig 109 Make a more permanent repair by removing the damaged section and replacing it with plastic pipe. Use a special adaptor to connect old to new.

Repairing Downpipes and Hoppers

If the pipe is rusted through or cracked because of accidental damage, you can make a temporary repair by wrapping a length of flashing tape around the pipe. Wire-brush the metal first to remove old paint, then brush on a coat of the flashing primer sold with the tape, and leave this to dry. Then cut a piece of tape to length, peel off the backing paper and wrap it tightly round the pipe. Paint it to match the rest of the pipe-run and disguise the repair. The quickest way of doing this is to use aerosol paint, masking the wall behind the downpipe with a piece of card.

Where rust or damage is extensive, it is best to replace the affected section completely. You can still buy cast-iron downpipe from builders' merchants if you want to replace like with like, but it is probably easier and cheaper to use plastic downpipe of the same diameter instead. Apart from new pipe and brackets, you will need a special adaptor to connect the end of the existing metal pipe to the new plastic one.

Fig 110 (*above*) Disconnect the gutter outlet from the top of the downpipe so you can feed in your hose and wash away the blockage.

Clearing Blocked Hoppers

Blocked hoppers need prompt attention for two reasons: the resulting overflow can be very messy, and smells may waft back into the bathroom. Start by scooping out debris with a rubber-gloved hand. Then scrub out the inside of the hopper thoroughly with washing soda to remove soap build-up from the sides and bottom. If the water still does not flow away, there may be a blockage in the downpipe leading from the hopper. Clear this as described opposite.

It is a good idea to fit hoppers with wire mesh covers to prevent debris from being blown into them, and to scrub them out at regular intervals to keep them clean.

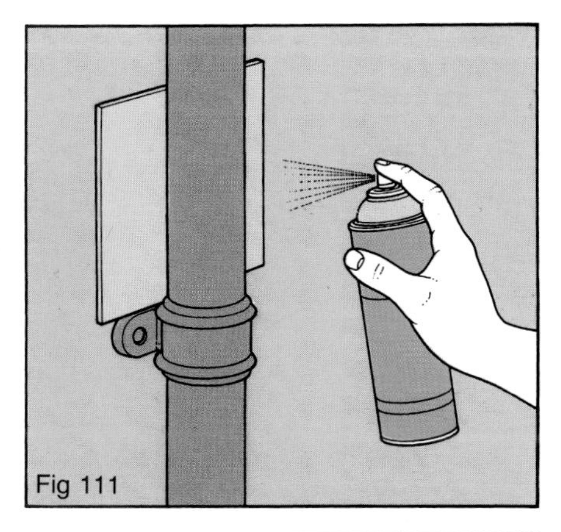

Fig 111

TIP
Fit a mesh balloon to gutter outlets to stop debris blowing or washing into downpipes. Similarly, fit mesh covers over hoppers.

Fig 112

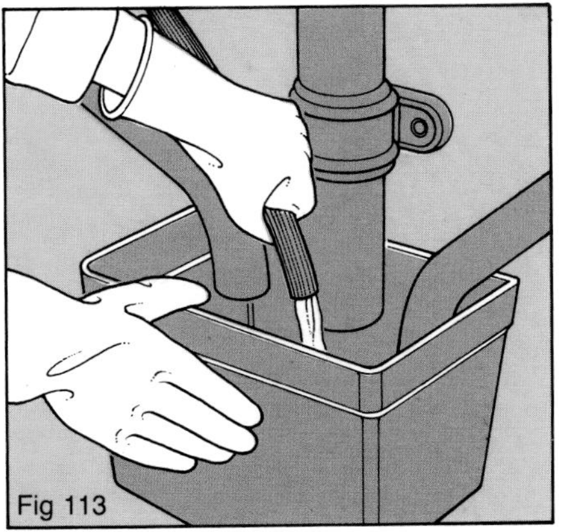

Fig 113

Fig 111 Paint downpipes with an aerosol paint, masking the wall behind with card as you work.

Fig 112 Clear debris by hand from a blocked hopper, then scrub out the interior with washing soda to remove built-up soap residues.

Fig 113 Flush the cleaned hopper through with plenty of fresh water.

Repairing Valley Gutters

The internal angle where two roof slopes meet, called a valley, may be formed of special interlaced or interlocking valley tiles on top-quality roofing work. However, on most roofs the tiles or slates are simply cut back at an angle and the junction between the two roof slopes is waterproofed with a metal-lined valley gutter into which water flows from the roof surface. This gutter discharges its contents into the eaves gutters or, where there is a parapet wall at eaves-level, into a rainwater hopper.

The gutter lining is usually made of lead, although zinc and aluminium alloy is also used. It can become porous with age, allowing water to seep through and cause rot in the roof timbers. High winds can lift and damage the metal sheeting, while blockages caused by moss and the like can cause overflows at the sides of the gutter. Again, water can then penetrate the roof itself and may cause rot in the timbers. Therefore, it is important to inspect valley gutters regularly, and to fix any faults that you find promptly.

What to do

Start by setting up your access equipment – ideally a platform tower erected in the angle where the two roofs meet so you have a safe platform at eaves-level from which to work. Because of the shape of the valley, you will not be able to use a roof ladder, so the eaves platform is essential for safety as you climb the valley itself. Wear shoes with rubber soles.

Clear any debris you find from the gutter, and check it for pin-holes, splits and lifted seams. If the surface is badly weathered, give the whole gutter two coats of bituminous waterproofer.

Where you find splits or tears, repair them with self-adhesive flashing tape. Wire-brush the area first, apply flashing primer and allow it to dry. Then cut a piece of tape to size, peel off the backing paper and press it into place. Run over it with a seam roller.

If more than one patch is needed, it is best to run a continuous length of tape right down the gutter from top to bottom.

What you need:

- access equipment
- bucket and trowel for collecting debris
- bituminous waterproofer
- wire-brush
- self-adhesive flashing tape and primer
- handyman's knife
- paintbrush
- wallpaper-seam roller

CHECK

- inside the roof space for signs of long-term water penetration and rot along the underside of the valley rafter. If you find rot, you may have to call in a professional to strip out the valley gutter so rotten wood beneath it can be treated or replaced as necessary
- that the mortar pointing beneath the cut edges of the tiles next to the valley gutter is intact, since loose pieces are a prime cause of blockages in the gutter. Repoint if necessary.

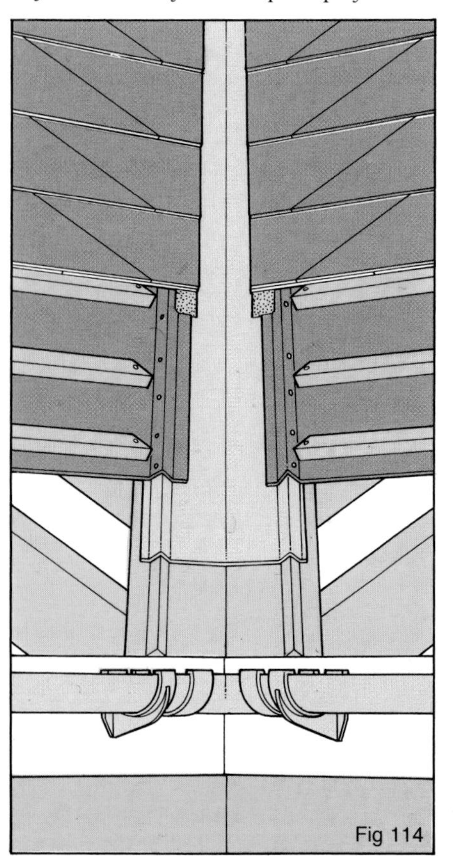

Fig 114 At a valley, a gutter board provides a solid base for the gutter lining. The roofing underfelt overlaps the lining down each side, and the tile edges are pointed to prevent water penetration.

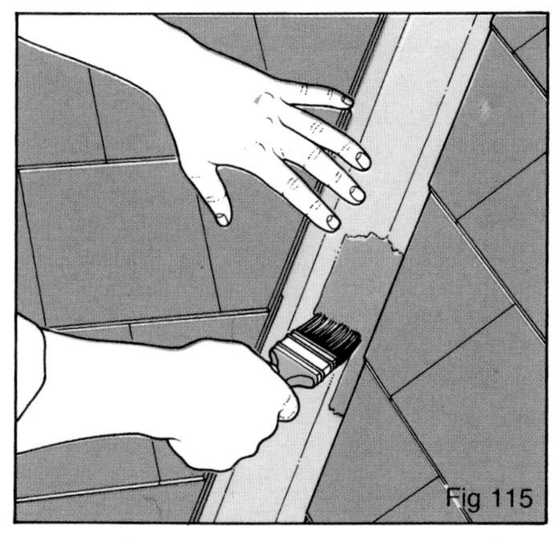

Fig 115 To patch a split or tear in the valley lining, wire-brush the area and then apply flashing primer.

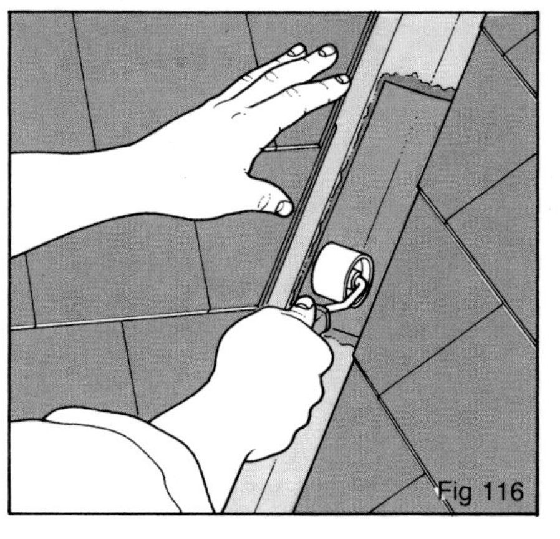

Fig 116 Cut a piece of flashing tape to size and press it into place. Apply extra pressure with a seam roller for a watertight bond.

WEATHERPROOFING JOBS

Your house roof and its rainwater disposal system obviously bear the brunt of heavy weather, but the house walls have to bear their fair share of the assault too, especially in exposed locations. There is just as much variety in the way walls are built up and down the country as there is in roof construction, although once again standardization of building practice has tended to lead to a depressing uniformity of style in recent years, as any new housing development will reveal. It may not necessarily result in a weatherproof wall.

Apart from the walls themselves, the major weak spots as far as weather penetration is concerned, are the openings formed in them for doors and windows. If these are not kept in good condition, it is all too easy for water to penetrate and for rot to take a hold of woodwork. Water can also penetrate around ill-fitting frames, leading to penetrating damp and disfigured decorations.

Most homes built in this country since the 1930s have walls built of cavity construction, with two 'skins' of masonry (called leaves) held together at intervals by wall ties bedded in the mortar courses. The cavity wall was introduced to help cut down on the incidence of water penetration that was such a problem with solid walls, especially if the bricks were at all porous or there were defects in the mortar pointing. By introducing a vertical cavity, any water penetrating the outer leaf of the wall could in theory run harmlessly down its inner face to ground level, leaving the inner leaf completely dry. However, in practice, poor workmanship can allow water penetration to reach the inner leaf for a variety of reasons.

One of the commonest is faulty positioning of the wall ties and carelessness in allowing mortar droppings to collect on them as the wall is built up. The result is to create bridges across which water can travel to the inner leaf and cause tell-tale damp patches.

Another common problem area is at door and window openings, where the cavity is bridged by the frames within the openings. Here, incorrect positioning of cavity trays across the head, and of damp-proof courses down the sides and across the sill, can again channel water to the inner leaf.

Fig 117 Plastic gutters and downpipes are virtually maintenance-free.

The third cause of trouble is the faulty installation of cavity-wall insulation. In an attempt to reduce the rate of heat loss through exterior walls, it seemed like a good idea to fill the wall cavity with insulating material. However, if the cavity is completely filled it interferes with the free run-off of water penetrating the outer leaf, and certain types of insulation – notably urea-formaldehyde foam – have been known to fissure if incorrectly installed and to allow water to run through to the inner leaf. The end result? Once again, penetrating damp.

The other major enemy of any house is rising damp, caused by moisture from the ground soaking upwards into walls and solid floors. It should not do so if damp-proof courses (DPCs) have been installed and are in good condition, but they may fail either locally or on a large scale and allow moisture to penetrate the structure. This not only ruins decorations; it can also lead to rot on floor timbers and is hardly good for the health of the house's occupants.

Putting right major faults like these is a job for the professional rather than the do-it-yourselfer. However, there are several smaller-scale weatherproofing jobs which can be easily undertaken by the home owner to keep his house in good condition.

Replacing Putty

One of the commonest reasons for water penetration is defective putty around windows. The purpose of the putty is to provide a bedding within the frames for the panes of glass, and also to act as a watertight seal but, as time goes by, the putty begins to crack and shrink away from the glass, allowing water to run down behind it. It can then not only seep in round the edges of the glass, but can also cause rot in the glazing bars or the frame and, if this gets a serious hold, the entire frame may need to be replaced. On metal frames, the water penetration can lead to rust and premature failure of the window.

Putty, especially the traditional linseed oil type, is in fact rather poor at its job, and, for this reason, many wooden windows now feature solid-timber glazing beads to retain the panes, which are bedded in either acrylic putty or non-setting mastic to provide a longer-lasting waterproof joint. It may be worth considering adopting this approach with existing timber-framed windows, especially in exposed locations.

Because of the relatively high failure rate of conventional putty, it is a good idea to inspect the outside of your windows at least once a year, in spring, for example, when the winter will have done its worst to them, and to note any areas of putty which are beginning to crack up. Even if the putty is sound, you may find that the paint film sealing its join with the glass has flaked off and should be replaced before water can begin to penetrate behind it.

Modern plastic replacement windows do not use putty or mastic at all; instead the glass (which is usually a double-glazed sealed unit) is retained in the frame by synthetic-rubber glazing beads.

What to do

If you find putty that is cracked, pulling away from the glass or missing altogether, the first job is to hack it away. You can use an old chisel for this, although you will get quicker results with a specialist tool called a glazier's hacking knife. This has a short stiff blade with a wedge-shaped cross-section, and the flat top edge can be struck with a hammer if necessary to remove stubborn bits of putty. As you hack out the old putty, you will come across small metal nails called glazing sprigs on wooden-framed windows, and metal clips on metal-framed ones. Leave these in place if you can; if they are rusty, remove them with pincers.

With all the old putty removed, check whether there is any sign of rot getting a foothold in wooden frames, and treat any suspect areas with wood preservative or a rot repair kit. Then prime any bare wood around the rebate with wood primer, and replace any rusty sprigs. With metal windows, use a wire-brush or wire wool to remove any rust spots and apply either a rust inhibitor or a one-coat anti-rust paint. Replace any missing glazing clips, which are fitted into small holes in the window frame.

You can now replace the putty, using an acrylic type in preference to linseed oil putty. Scoop some out of the container and form it into a neat ball in your hand. Then squeeze it out between your finger and thumb, and press it firmly into the rebate all round the window. Next, use a putty knife drawn along at an angle to form the putty into a neat triangular bead with a smooth surface, with neatly mitred corners. Finish off the outside by drawing a dry paintbrush along the join between putty and glass to seal it.

What you need:

- old chisel or glazier's hacking knife
- hammer
- pincers
- wood preservative or rot repair kit
- wood primer
- rust treatment
- paintbrush
- replacement sprigs or glazing clips
- acrylic putty
- putty knife
- filling knife
- glazing beads
- fixing pins
- non-setting mastic

CHECK

- that wood windows are free from rot before replacing the putty. Treat suspect areas with wood preservative or a rot repair kit
- that metal windows are free from rust. Wire-brush the surface and treat it with a rust inhibitor, or use rust-resistant paint.

TIP

Keep putty indoors before using it in cold weather, to make it softer and easier to handle.

Fig 118 Hack out all defective putty with an old wood chisel or a glazier's hacking knife. Remove any rusty sprigs or glazing clips.

Fig 119 Brush wood primer on to any bare patches around the rebate.

Replacing Putty

If water has been seeping in around the glass, use a filling knife to rake out as much of the old bedding putty as you can on the inside. Then press new putty in all round, and trim off the excess to leave a neat, smooth finish.

Leave the new putty to harden for about fourteen days. Then paint over it to protect it and to seal the join between putty and glass – the paint should extend onto the glass by about 3mm (⅛in). Use masking tape to ensure straight edges.

Using Glazing Beads

Remove the old putty as before, and prise out the old glazing sprigs. Then cut the beading to length to fit the four sides of the pane, and apply putty or non-setting mastic along the two rear edges of each length. Press the length into place so the putty/mastic forms a continuous seal against the glass and the rebate, and drive in the fixing pins. Repeat this for the other lengths, then trim off excess mastic all round. Finally, paint or stain the beading to match the frame.

Fig 120 (*above*) Always paint over new putty after about fourteen days, taking the paint on to the glass by about 3mm to seal the glass/putty joint.

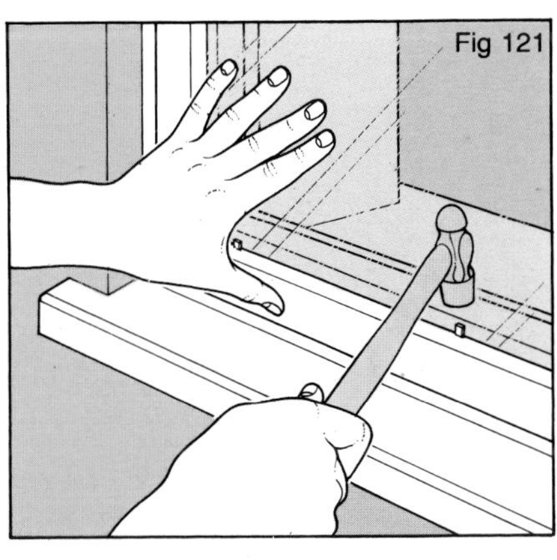

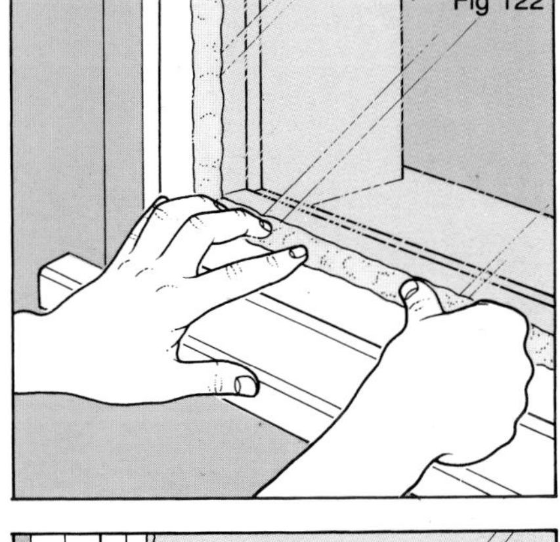

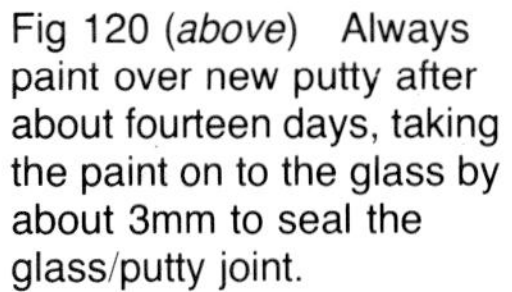

Fig 121 Replace any sprigs or glazing clips that you removed earlier.

Fig 122 Press the new facing putty firmly into place with your thumb all round the rebate.

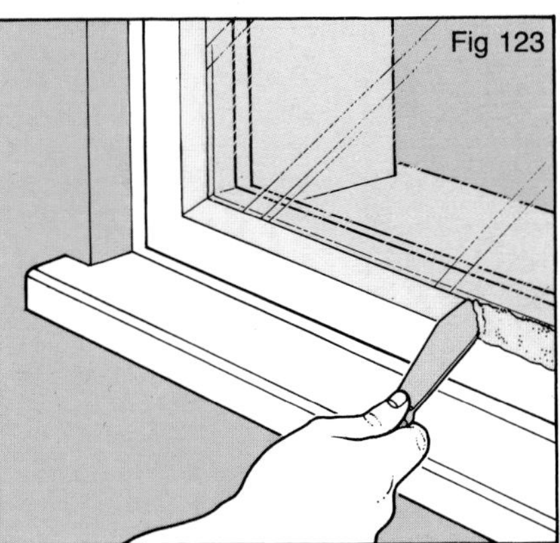

Fig 123 Use a putty knife to form a neat bevel all round, and to leave neatly mitred corners.

Fig 124 If water has penetrated inside, rake out as much of the old bedding putty as possible and press new putty in all round. Trim off the excess for a neat finish.

Repairing Leaded Lights

Leaded lights have been around for hundreds of years, from the days when good clear glass could be produced only in small pieces and the lead lattice was developed to allow larger window areas to be glazed. They were particularly popular during the 1930s, when many a semi-detached also had its own individual area of stained glass on the front door and at the tops of the bay windows. Nowadays they are available once again, but in the form of double-glazed sealed units with the lead strips actually bonded to the face of a single sheet of glass. They look similar, but they are somehow too perfect to deceive anyone.

The main problems with traditional leaded light windows are deformation of the whole pane (often due to over-zealous window cleaners applying too much pressure to the centre of the pane) and leaks around individual lights. Each is held in a channel in the edge of the lead lattice strips (called cames), and is bedded in putty, which can crack and fall out as time goes by, allowing water to penetrate.

What to do

If the whole pane has become distorted, it is best to remove it completely from its frame so it can be repaired on the workbench. Board up the window opening left by its removal, or cover it with heavy-duty polythene.

Lay the pane on the bench, bulge uppermost, and then place an offcut of plywood or similar board over it. Press downwards carefully but firmly to force the lattice back flat.

Next you can attend to the leaks. Start by soldering any joints that have popped as a result of flattening the pane. Then work putty (ordinary metal-casement putty coloured with powder paint) around the edges of leaking lights, on both faces of the pane. Next, press down the edges of the came around each light in turn with a blunt tool, and trim off any excess putty with a putty knife. Finally, burnish up the cames with a wire brush, and replace the repaired pane in its frame.

As an alternative to putty, you can seal minor leaks with clear silicone mastic.

What you need:

- general tools for removing pane
- plywood panel
- soldering iron
- wire solder
- metal casement putty plus powder paint
- putty knife
- wire-brush

CHECK

- that all the lights are intact before attempting to flatten the pane. If any are cracked, they may break completely

TIP

For a temporary repair, seal cracks in individual lights by brushing a little clear polyurethane varnish along each side of the crack with an artist's paintbrush.

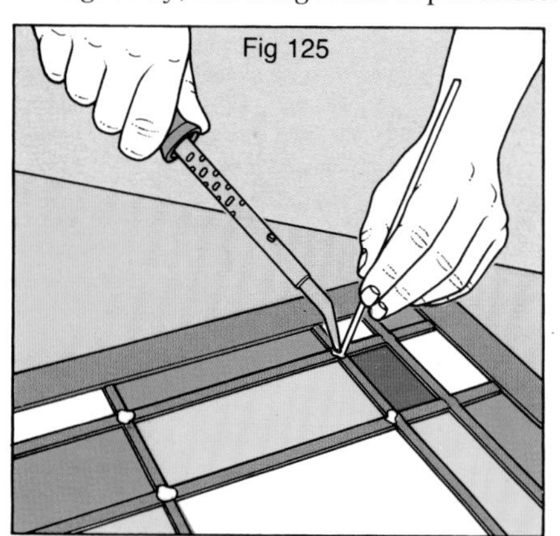

Fig 125 If possible, always remove the complete pane from its frame to carry out repairs. If joints have opened up, press them closed and solder them using wire solder and a soldering iron.

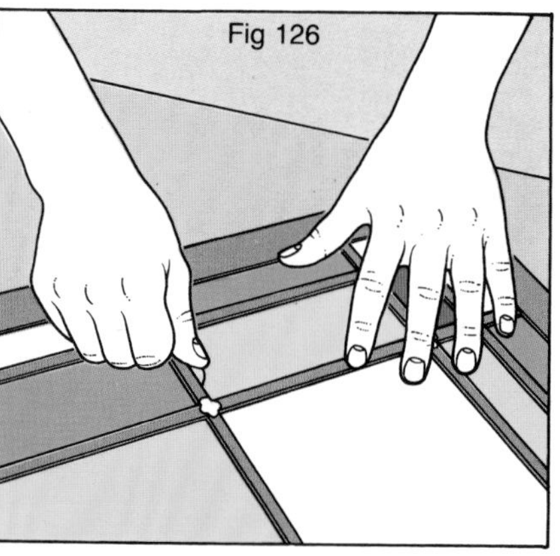

Fig 126 Press coloured metal casement putty in around all edges and both sides of leaking lights.

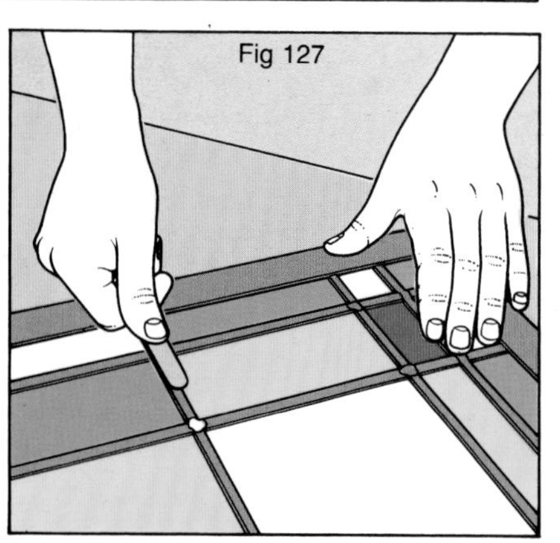

Fig 127 Then press the cames down onto the glass with a blunt instrument, and trim off any excess putty.

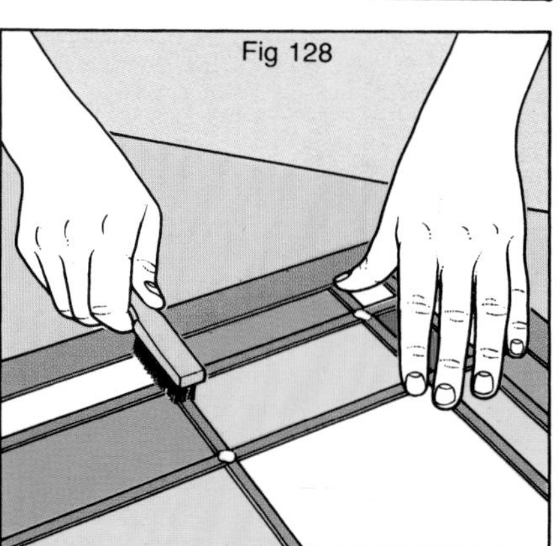

Fig 128 Finally brush up the cames with a wire-brush, and replace the pane in its frame.

Sealing Around Frames

Most windows and door frames are set directly within openings formed in the house walls; others, mainly metal-framed types, often fit within a timber sub-frame. On relatively recent homes, there will be vertical and horizontal damp-proof courses (DPCs) around the frame to prevent rain from being driven in between the frame or sub-frame and the masonry, but if these were incorrectly installed, damaged during the building work (a common occurrence), or are simply missing altogether on an older property, then rain penetration can take place. The result is areas of damp around the openings, often spreading on to surrounding walls.

Often attempts are made to seal this weak spot by using mortar or exterior-quality filler, but these rarely work for long because differential movement between the frames and the masonry always occurs as temperature and humidity conditions vary, and this movement simply causes rigid fillers like these to crack and fall out, allowing further rain penetration to take place.

What to do

The first step in curing the problem is to rake out any old filler that has been used to stop up the gaps between frame and masonry. Use an old screwdriver or similar tool for this, then brush out all loose debris with an old paintbrush.

The best material to use for filling gaps up to about 10mm (3/8in) wide is a non-setting exterior-quality mastic. This is sold in cartridges with fit in a standard cartridge gun; squeezing the trigger forces a piston up against the base of the cartridge and extrudes the mastic from the nozzle. Cut this with a sharp knife so that the bead is slightly larger than the gap to be filled, and simply pipe it into the gap. Wipe over the bead with a wet finger to ensure good adhesion and a smooth finish.

For larger gaps, it is better to use an aerosol filler foam which expands as it is extruded from the container. Pipe it into the gap and leave it to harden. Then use a sharp handyman's knife to trim off any excess foam, and paint over it to disguise the repair.

What you need:

- old screwdriver
- old paintbrush
- mastic cartridge
- cartridge gun
- aerosol filler foam
- sharp handyman's knife

CHECK

- that the nozzle is clear on a partly used cartridge; material left in the nozzle will set semi-rigid, and will have to be poked out before more mastic can be extruded
- that you read the instructions on cans of aerosol filler foam carefully, and follow them to the letter, particularly as far as safety instructions are concerned

Fig 129 Start by raking out all old rigid filler from round the frame, using an old screwdriver or similar tool.

Fig 130 If you are using mastic, cut the nozzle to give a bead a little wider than the gap to be filled, and pipe this into place. Wipe over it with a moistened finger for a smooth finish.

Fig 131 If you are using filler foam, pipe it into the gap and leave it to set hard.

Fig 132 Then trim off any excess with a sharp handyman's knife.

The majority of homes in this country have exterior walls of brick. How well they stand up to the passage of time depends on a number of factors, including the correct choice of brick and mortar mix, the right profile for the pointing, and on other factors such as the presence of salt spray in coastal areas or chemical pollution in industrial ones.

The wrong type of brick or mortar mix, coupled with an inappropriate pointing profile, can allow rainwater to penetrate behind the face of the bricks; in cold weather it then freezes and bursts the face of the brick – a fault known as spalling. Once the brick face is damaged, further water penetration takes place and extends the damage, while on solid walls it can also cause damp patches on inside walls.

Adverse atmospheric conditions can also gradually erode the faces of the bricks. This is not only unsightly; it can again allow moisture penetration, and if other faults are also present then serious damp can result. The solution is to inspect your brickwork regularly for faults.

Using Masonry Sealers

If your exterior walls are sound but you suffer from penetrating damp and cannot locate a specific cause, it may be that the bricks themselves are porous and are allowing water to soak in. A relatively simple and inexpensive way of tackling this is to treat the whole wall surface with a liquid silicone water-repellent sealer.

Clean down the brickwork thoroughly first of all, using fungicide to kill any algal growth on the surface. Then brush the sealer liberally into the surface and leave it to dry.

What to do

If your inspection reveals any bricks that have spalled as a result of frost damage, the only long-term remedy is to chop them out and replace them with new matching bricks, mortared into place.

To make it easier to remove the brick without damaging its neighbours, start by drilling a series of holes into the face of the brick with a masonry drill bit and a power drill. Fit a depth stop to match the depth of the brick (102mm/4in), so you will know when you have drilled right through it.

Next, use a sharp cold chisel and a club hammer to chop out the honeycombed brick bit by bit, and also the old mortar surrounding it. On cavity walls, try not to allow broken pieces of brick to fall into the cavity, where they could lodge and form a potential damp bridge.

When you have completely cleared out the recess, mix up a small quantity of 1:5 cement:sand mortar (or buy a small bag of dry ready-mixed repointing mortar).

What you need:
- fungicide solution
- clear-silicone masonry sealer
- paintbrush
- electric drill and masonry bit
- club hammer
- cold chisel
- pointing trowel
- 1:5 cement:sand mortar
- brick bolster

CHECK
- that pieces of broken brick do not fall into wall cavities
- that you do not damage the DPC if working on bricks near ground level
- that the new pointing matches the old in colour, by mixing up a small batch and leaving it to dry. Add mortar pigment if necessary

Fig 133 Apply clear silicone sealer to porous brickwork to prevent damp penetration.

Fig 134 To remove a spalled brick, first drill a series of holes in it using a masonry bit in your power drill.

Fig 135 Then chip out the honeycombed brick and the surrounding mortar with a cold chisel and club hammer.

Trowel a bed of it into the bottom of the recess, and butter more mortar on to the ends and top surface of the replacement brick. Offer it up to the recess and carefully tap it into position. Scrape off any excess mortar which oozes out of the joints, and point up the joints neatly all round the brick to complete the repair.

If you have difficulty in chopping out the whole brick, try to remove at least half its thickness by drilling and chiselling. Then, place mortar in the base of the recess as before, cut the replacement brick in half lengthways, and butter mortar on to its top, ends and rear face. Tap it into the recess so its face finishes flush with the surrounding brickwork, and repoint as before.

If your inspection of your outside walls shows that the bricks are still sound but the pointing is beginning to deteriorate, now is the time to chisel out defective areas and to repoint them with fresh mortar. To minimize water penetration, especially in exposed areas, use flush or weatherstruck pointing rather than any of the recessed profiles.

Fig 136 (*above*) You can hire a small hand-held damp meter to check interior walls for signs of penetrating damp.

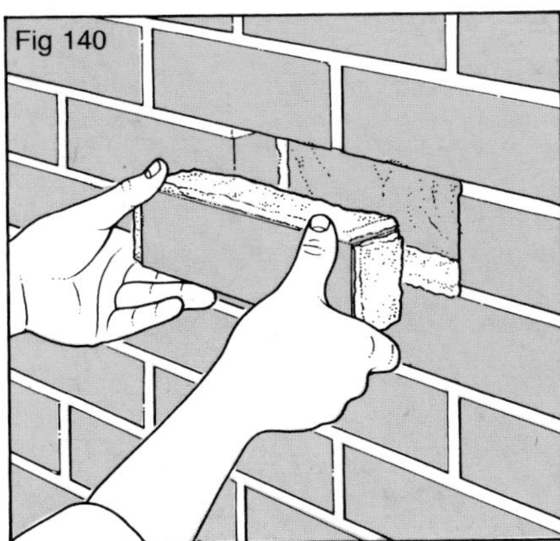

Fig 137 Trowel some 1:5 cement:sand mortar into the bottom of the recess.

Fig 138 Butter mortar on to the ends and top of the replacement brick and tap it gently into place.

Fig 139 Make sure the brick face is flush with its neighbours, and then point around it neatly.

Fig 140 If you cannot remove the whole brick, chop out at least half of it and then mortar in a new half-brick, split lengthways.

Clearing DPC Bridges

Most homes built since the late nineteenth century have a physical damp barrier known as the damp-proof course (DPC) built into the exterior walls to stop ground moisture from being drawn upwards into the masonry. The DPC can fail locally, especially where materials such as slate or engineering bricks were used, allowing tell-tale semi-circular damp patches to appear low down on the inside of the wall. If this occurs, a new DPC can be injected into the masonry (*see* page 88) to cure the problem.

However, if you spot rising damp, do not immediately suspect DPC failure; the cause may be something bridging the DPC and allowing moisture to bypass it.

Some of the commonest DPC bridges are shown below. The first is caused by a wall or building being erected next to the house without either a horizontal or vertical DPC being included. The second results from soil or other material being piled against the wall above DPC level. The third is due to a step or other slab being built to a level higher than the house DPC without a vertical DPC between wall and slab, again allowing moisture to bypass the house DPC. The fourth is the result of rendering being applied to the wall over the DPC line.

Lastly, damp can also occur if hard surfaces are built close to DPC level and rainwater splashes up the wall above it.

What to do

Either remove the DPC bridge, or incorporate an appropriately sited DPC to prevent damp from travelling above DPC level – *see* below.

What you need:

- strip DPC material
- brush-on bituminous damp-proofer
- paintbrush
- sundry building tools

CHECK

- that path and patio levels adjoining the house are kept at least 150mm (6in) below DPC level, so rainwater falling on them cannot splash up onto the wall above the DPC

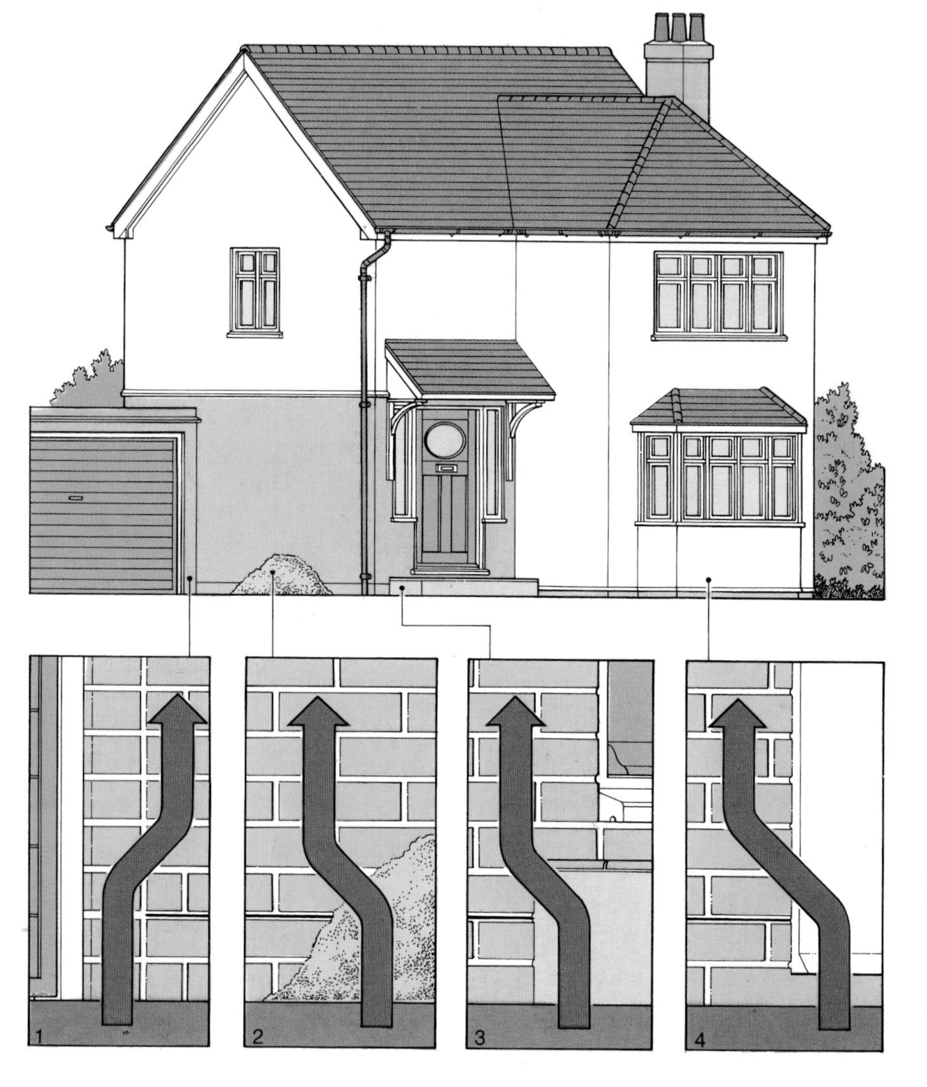

Fig 141 Four common DPC bridges. In **1**, either insert a horizontal DPC in the wall adjoining the house, or fit a vertical DPC between the two walls. In **2**, remove the soil heap. In **3**, fit a vertical DPC between slab and wall. In **4**, cut back the rendering to just above the level of the DPC.

INSULATION JOBS

Insulation means saving energy, on every level, from the personal to the global. As people become increasingly conscious of the importance of environmental issues, one of the greatest contributions everyone can make is to cut down on unnecessary waste of fossil fuels, and to reduce the amount of carbon dioxide released into the atmosphere by burning them. This means making more efficient use of energy, and insulation has a big part to play in this. It will save you money too.

Insulation is a means of stopping heat transfer from a warm area to a cold one. In this country, the outside air temperature is below what most people regard as a comfortable figure for much of the year, so we need to heat our homes up to compensate for this.

The problem is that all the materials we traditionally use in house-building conduct heat to a greater or lesser extent. Wood is a fairly good insulator, brick an average one and glass is downright poor, as anyone who has sat next to a window on a cold winter's day will testify.

Worst of all, providing well-insulated homes has until comparatively recently been a very low priority, both for housebuilders and for the legislators who frame the Building Regulations with which builders must comply.

However, at last the tide is turning, and the latest amendments to the Building Regulations require much higher standards of insulation than ever before for new buildings. They have also belatedly recognized the problems that over-insulation can cause as far as condensation is concerned, both inside the house itself and also within the building's structure. This means that new houses may cost a little more, but will consume up to 20 per cent less heat.

Unfortunately, this will not help those people living in older properties, many of which were originally built with no thought to their insulation performance at all. Of course, over the years various attempts will have been made to insulate houses like these, but what was deemed adequate ten or twenty years ago is well below par nowadays. Therefore it will certainly pay for you to check out your existing insulation, with a view to improving it for the future.

Fig 142 (*above*) Insulating tanks and pipework in the loft is an extremely cost-effective job to tackle.

Cost-Effectiveness

Before thinking about individual types of insulation, it is important to understand the concept of cost-effectiveness. Insulation costs money to install and can benefit you in either of two ways. It can enable you to reduce your heating bills, since your home will waste less heat and you can maintain the same internal temperatures without burning so much fuel. The annual saving that you make as a result will 'pay back' the cost of the extra insulation.

If you are also considering replacing your heating system, having better standards of insulation may also mean you can specify a less powerful (and less expensive) boiler and smaller (or fewer) radiators, which will be an indirect one-off saving, but valuable nonetheless.

Alternatively, you can enjoy higher internal temperatures than before without increasing your heating bills, but in this case there will be no direct savings, just a better degree of comfort.

As a guide to cost-effectiveness, loft and hot-tank insulation and draughtproofing score the highest, while double glazing and professionally installed insulation are the least cost-effective. Other insulation measures fall somewhere between the two.

Laying Blanket Loft Insulation

If you have a pitched roof, and you use the loft space just for storage, insulating the loft floor is one of the easiest and cheapest improvements you can carry out. You do this by filling the spaces between the joists with insulation material, either glass fibre or mineral wool, sold as blankets by the roll or in slab form, or else loose-fill material (vermiculite, a lightweight expanded mineral, or shredded mineral wool fibres), sold in bags. You can also have loose-fill insulation, usually mineral wool or fireproofed cellulose fibres, blown into the loft by specialist contractors (*see* page 85 for more details).

Blanket and slab materials are generally easier to handle than loose-fill types unless your loft is awkwardly shaped, contains a lot of obstructions or has irregular joist spacings. Whichever type you choose, you should aim to lay it to a depth of 100mm if you already have some fairly recent loft insulation, laying it on top of the existing material. Increase the depth to 150mm (6in) in uninsulated lofts, or those with just the old and woefully inadequate 1in thickness demanded by the 1965 Building Regulations.

Apart from being awkward to handle, loose-fill materials have another drawback. To be as effective as blanket types, they need laying to a greater depth, at least an extra 25mm (1in). With few ceiling joists being deeper than about 150mm (6in), there is nothing to contain the insulation unless you are prepared to fix battens along the top edge of every joist. Therefore, check your joist depth before using this type.

What to do

Before you start to actually lay the insulation, there are several things to check or to attend to. The first is to clear the loft of stored items so you have room to work; the best practical solution is to stack everything in one half of the loft, lay the insulation, then move the stack to the insulated end while you tackle the rest of the loft.

The second thing to check is that you have adequate loft ventilation (*see* page 59

What you need:

- vacuum cleaner
- mastic and cartridge gun
- cable clips
- hammer
- blanket, slab or loose-fill insulation
- polythene sheeting for use as vapour-check
- staple gun or adhesive tape
- kneeling board
- levelling batten
- gloves
- face-mask

TIP

Many people find that insulation materials irritate their skin, and loose fibres or dust can also be a problem. Always wear gloves and a disposable face-mask when laying them.

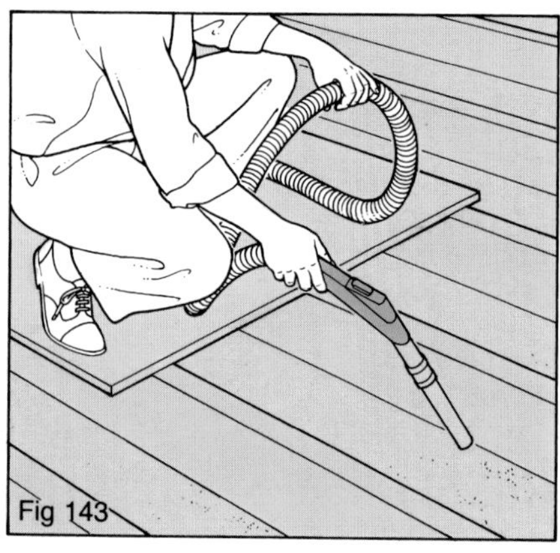

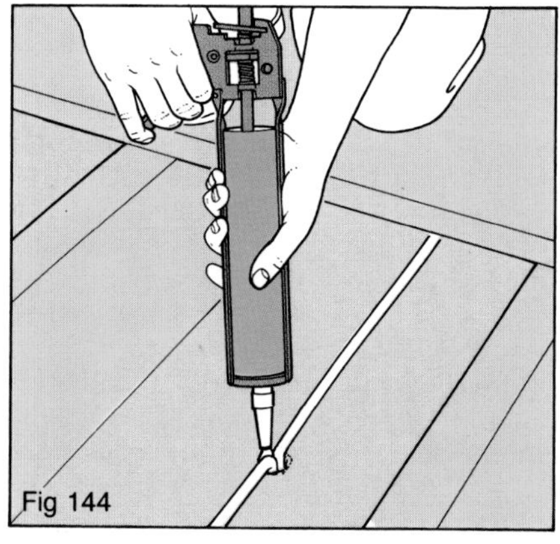

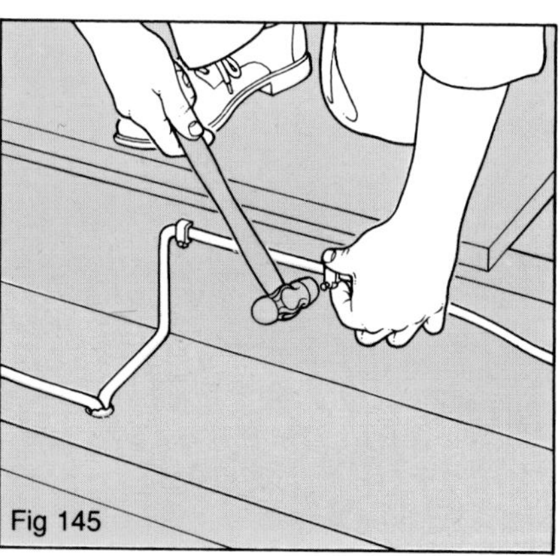

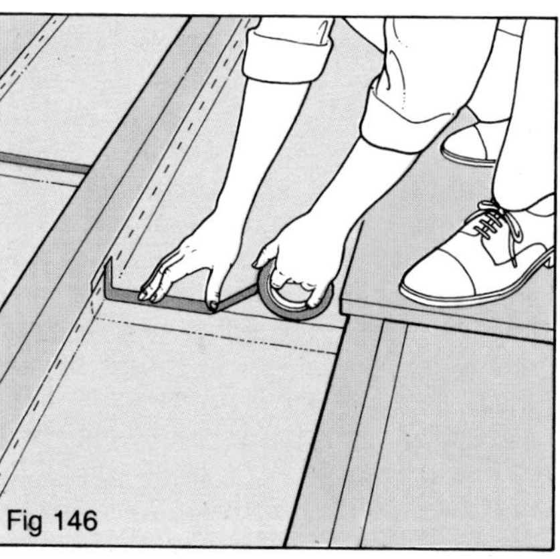

Fig 143 If there is no loft insulation present at all, use a heavy-duty vacuum cleaner to remove dust and debris from the ceiling surface.

Fig 144 Fill the holes and gaps through which cables and pipes pass, using non-setting mastic or aerosol filler foam.

Fig 145 If possible, lift cables off the ceiling surface and clip them to the sides of the joists.

Fig 146 To prevent condensation within the loft caused by moisture vapour rising from rooms below, use a staple gun to fix polythene sheeting between the joists. Tape the overlaps.

for more details). Next, to minimize the incidence of condensation in the loft, you should block up any gaps and cracks in the ceiling below which could allow warm, moist air into the loft space. It is a good idea to vapour-proof the bathroom ceiling, either by painting it with a solvent-based paint, or by laying polythene sheeting on the floor of the loft above it before you put down the insulation.

With blanket types, start work at the far side of the loft. Place the roll near the eaves and unroll the end, pushing it into the eaves as far as the wallplate on which the ceiling joists rest. Then roll out the blanket, tucking it lightly down between the joists but not compressing its thickness unduly. At the other side of the loft, simply tear the blanket across with your hands to the length you want. When one roll runs out, butt-joint the next up to it, and carry on as before.

With slabs, use a kitchen knife to cut each one down to match your joist spacing, and then simply lay the slabs between the joists, butting them tightly together. The

Fig 147 (*above*) Insulation slabs are more manageable than blankets but are only 50mm (2in) thick, so you need two layers.

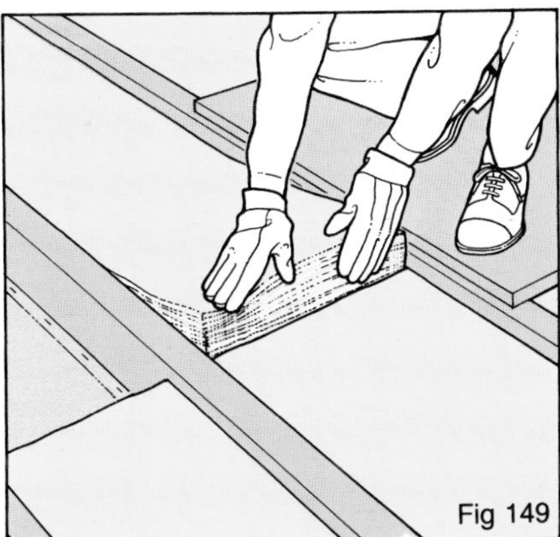

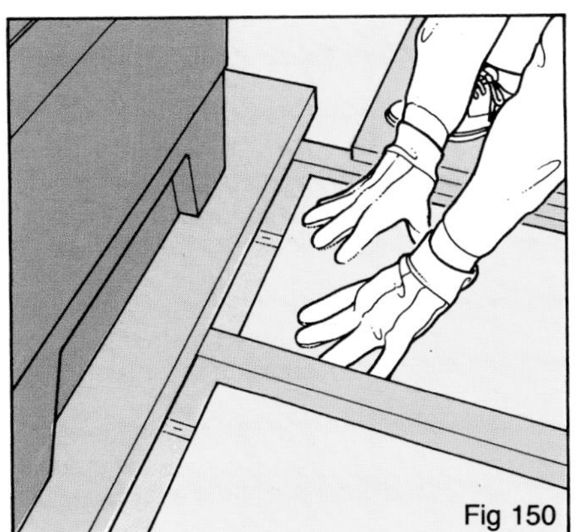

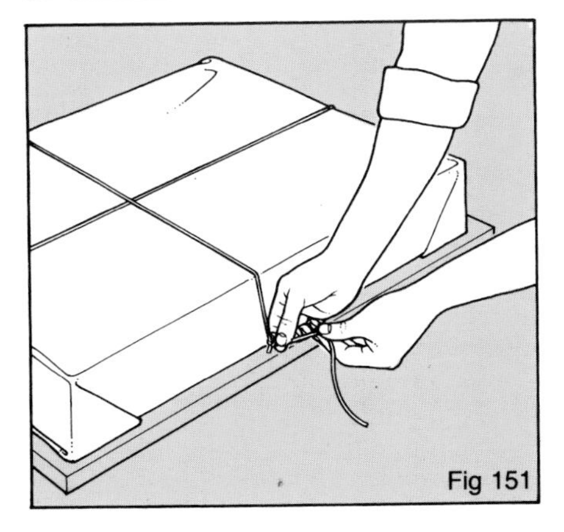

Fig 148 With blankets, position the end of the roll at the eaves and unroll the blanket back across the loft floor. Tuck the insulation down lightly between the joists as you work.

Fig 149 Form a tight butt joint between the ends of adjacent lengths.

Fig 150 Do not lay insulation beneath cold-water storage tanks or central-heating header tanks.

Fig 151 Wrap pieces of blanket in polythene and tape the package to the upper surface of the loft hatch-door.

Laying Loose-Fill Loft Insulation

slabs are 50mm thick, so add a second (or even a third) layer on top of the first to get the thickness you need.

With loose-fill material, simply pour the insulation out on to the loft floor and spread it evenly between the joists. Fluff mineral wool types up with your fingers; use a spreader to level off granular types.

As you lay the insulation, lift wiring above it wherever possible to prevent the risk of the cables overheating underneath it. If polystyrene granules have been used in the past to insulate the loft floor, it is a good idea to move them from any areas where cables are present, since contact with the granules can cause eventual deterioration of the cable sheathing.

When you have completely insulated the loft floor, secure a piece of insulation blanket to the top surface of the loft hatch, and draught-proof the rebate on to which it closes. Remember to insulate the cold-water storage tank, the feed-and-expansion tank that tops up your heating system, and also any exposed pipework within the loft (*see* pages 68–71).

Fig 152 (*above*) Loose-fill material is easier to handle than blankets or slabs in lofts with obstructions.

Fig 153 If you are laying loose-fill material, nail scrap wood between the joists to stop it from spreading under water tanks.

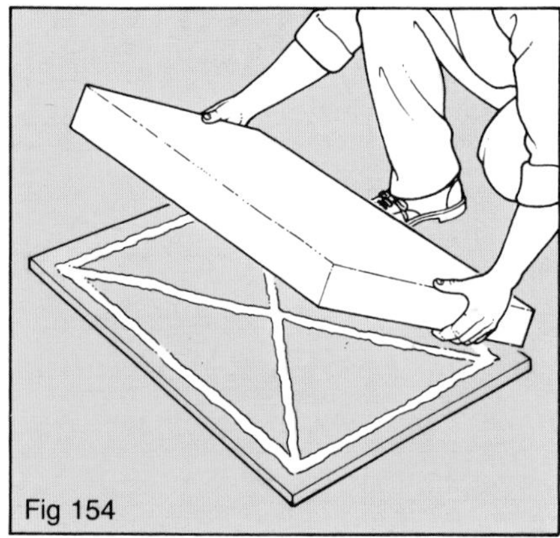

Fig 154 Stick a slab of rigid polystyrene foam to the top of the loft hatch-door.

Fig 155 Simply pour loose-fill materials out between the joists.

Fig 156 Use a spreader to level out granular materials. Fluff fibrous types up with your fingers first.

Improving Loft Ventilation

As the importance of loft insulation became apparent during the 1960s, installation instructions exhorted householders to stuff the blankets well into the eaves to eliminate draughts. Thousands of householders did so, only to discover years later that doing so had caused totally unforeseen problems with rot in their roof timbers. Because the natural ventilation of the loft space had been cut down by stuffing insulation into the eaves, the warm, moisture-laden air rising into the loft from the house below was trapped. Fitting the insulation had made the loft space considerably colder, so the moisture vapour was condensing to water on the roof timbers (and also on the insulation, saturating it in many cases and so making it useless as insulation). There was no ventilation to dry the timbers out, and it did not take long for rot spores to begin to develop, leading in some cases to serious damage to the roof structure.

It is now realized that good ventilation is essential for all roof spaces, and the current Building Regulations contain detailed provisions for ensuring that all new pitched and flat roofs are adequately ventilated. Existing roofs should be checked to ensure that they are ventilated to the same standard if rot problems are to be avoided.

What to do

With pitched roofs, ensure that insulation is pulled back clear of the eaves. It is a good idea to install small V-shaped eaves vents at opposite sides of the roof, to hold the insulation away from the underside of the roof slope and allow air to circulate over the timbers.

In addition, you should create ventilation openings in the underside of overhanging eaves, either by drilling holes in the soffit with a wood bit and covering them with fine mesh to prevent insects from entering, or by making larger cutouts so you can fit proprietary slot ventilators into the holes. It is also a good idea to install air-bricks in gable walls.

What you need:
- eaves vents
- electric drill with wood bit or
- electric jigsaw
- wire mesh or proprietary eaves ventilators
- air-bricks
- masonry tools

CHECK
- that pitched roofs have eaves ventilation openings equivalent in area to a continuous slot 10mm (3/8in) wide, running the full length of the eaves

TIP
Ensure that recessed lights set in the loft floor have adequate ventilation by building an open box round them to hold back the insulation.

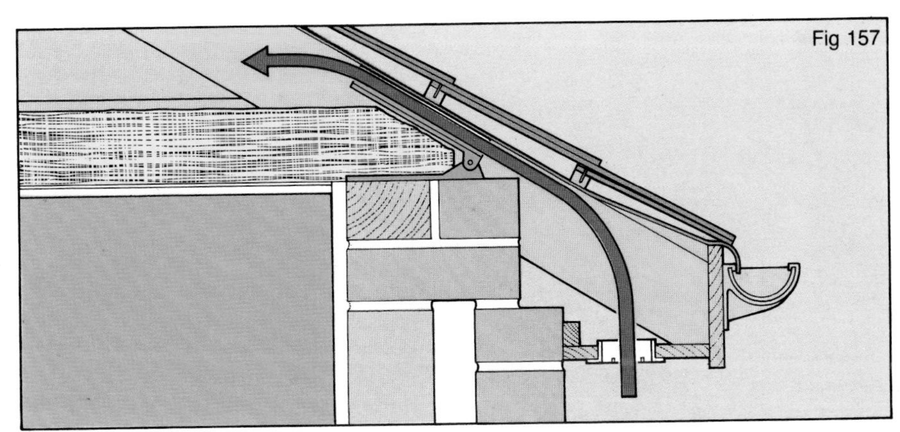

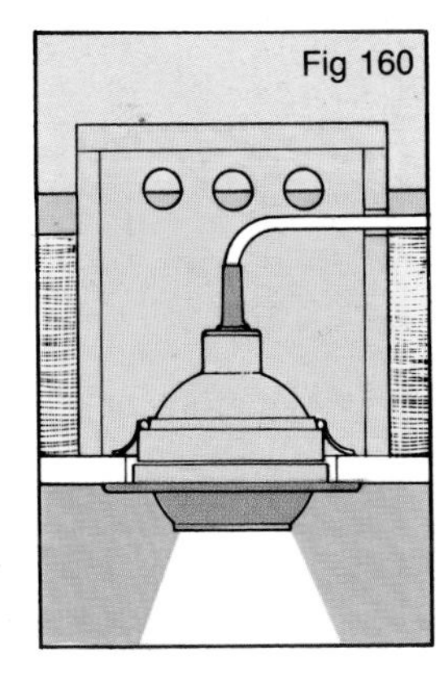

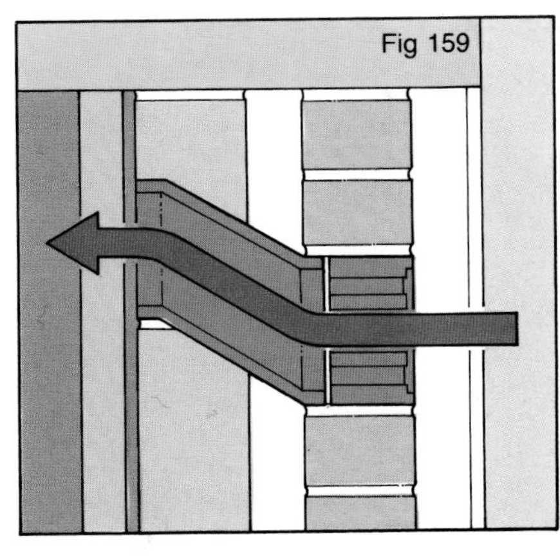

Fig 157 Fit eaves and soffit vents to opposite sides of the roof to ensure adequate ventilation.

Fig 158 Make holes or slots in soffits and fit mesh or proprietary vents.

Fig 159 Install air-bricks in gable walls to improve air flow in the loft.

Fig 160 Keep loft insulation clear of recessed lights.

Insulating Roof Slopes and Loft Rooms

If you have rooms in your loft, or plan to have a loft conversion at some future date, it is obviously a waste of effort and materials to insulate the loft floor. Instead, you should insulate the underside of the roof slope and the walls of the loft rooms themselves, so that you can benefit from heat rising into the loft from the house below.

You can insulate the roof slope in an unconverted loft in one of three ways: by fitting loft insulation blanket, insulation batts or rigid polystyrene boards between the rafters; by stapling paper-faced insulation mat to the underside of the rafters; or by fixing insulating plasterboard directly to them. If you use insulation blanket, batts or boards, you need a polythene vapour barrier on the room side of the insulation to prevent moisture vapour passing into the cavity behind it and possibly causing rot in the roof timbers. You then have to fix up sheets of plasterboard to form the walls and ceiling of the loft rooms. Insulating plasterboard, which can also be used to line the walls of the new loft rooms, contains a vapour barrier between the board and the insulant, and so kills two birds with one stone.

The vital point to remember when insulating a roof slope is that ventilation of the roof slope timbers – the rafters and tiling battens – is just as important as having good air flow over eaves timbers. This means leaving an air gap 50mm (2in) wide between the insulation and the underside of the tiles, slates or roof underfelt open at eaves level.

If you already have a loft conversion, adding insulation is more difficult, especially above sloping ceilings. Here, so long as you have access to the eaves, you may be able to slide cut-to-width slabs of rigid polystyrene up into the space between the room ceiling and the roof slope, and also to insulate the room walls from the loft side with slabs of polystyrene set between the wall studs. If you do not have access to the eaves spaces, the only practical solution is to line the walls and ceilings of loft rooms with insulating plasterboard.

What you need:

- insulating blanket, batts or polystyrene boards
- polythene vapour barrier
- plasterboard
- staple gun
- fixing nails
- battens
- fine-toothed saw
- hammer
- gloves
- face-mask

CHECK

- that you maintain the required 50mm (2in) air gap between the insulation and the underside of the roof slope
- that insulation is kept clear of the eaves
- that successive lengths of vapour barrier have a 50mm overlap

TIP

If you are using insulating plasterboard, note that you will need fixing nails long enough to pass through the board and the insulation behind it.

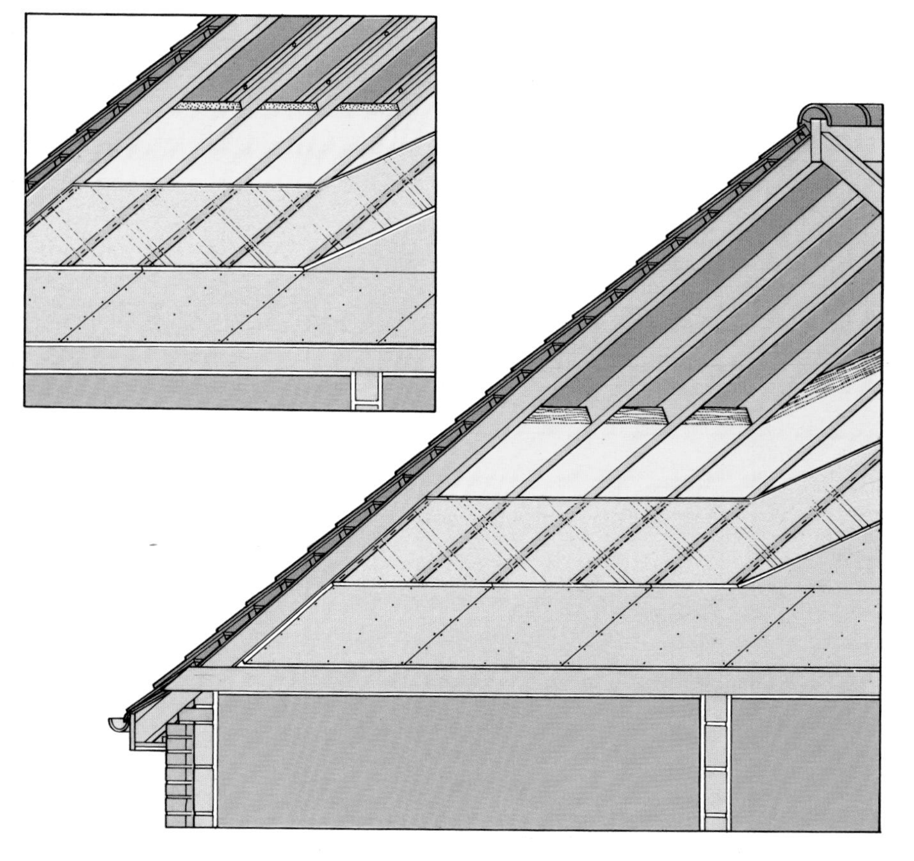

Fig 161 Insulate the underside of a roof slope by placing blanket or batts between the rafters. Add a vapour barrier, then fix up plasterboard. Alternatively, use rigid polystyrene boards held against battens pinned to the rafter sides.

Insulating Roof Slopes and Loft Rooms

What to do

If you plan to use insulation blankets or batts to insulate the underside of the roof slope, the most important thing is to maintain the 50mm air gap between the insulation and the underside of the roof slope. So choose a thickness of material that is at least 50mm less than the rafter size (50mm thick for 100 × 50mm rafters, for example).

With semi-rigid batts, it is usually possible to wedge the slabs firmly in place between the rafters, and then to fit the polythene vapour barrier when all the insulation has been placed. However, with the more flexible blanket types, it is simpler to work downwards from the ridge and to fix insulation and vapour barrier at the same time. Start by securing the top end of the first length of blanket to the side of the ridge, unroll it down to the eaves and cut it to length. Fix a second and third length alongside it in adjacent bays. Then staple the top end of the vapour barrier to the ridge board, and work down the rafters towards the eaves, stapling the vapour barrier to the rafters to hold the insulation blanket in place between them.

Paper-faced insulation is easier to handle than plain blanket, since it has paper flanges at each side which you staple to the undersides of the rafters. You again need a vapour barrier over it.

With polystyrene boards, careful cutting and regularly spaced rafters may allow you to wedge individual slabs in place between the rafters. However, it may be quicker to staple slim battens to the sides of the rafters first, and then to secure the slabs in place against the battens with nails tapped into the rafter sides. Then, you can staple the vapour barrier in place as before.

Fig 162 (*above*) Insulation batts are rigid enough to stay in position when placed between the rafters.

Insulating Room Walls

You can use the same techniques for insulating the walls of rooms built in the loft, by placing blankets, batts or polystyrene boards between the timber studs forming the walls. Make sure a vapour barrier is included on the room side of the insulation, and check that insulation is lifted from the loft floor as the rooms are constructed.

Alternatively, you can line the ceilings and walls of loft rooms with insulating plasterboard. Cut individual boards to size, then nail them directly to the room's timber framework.

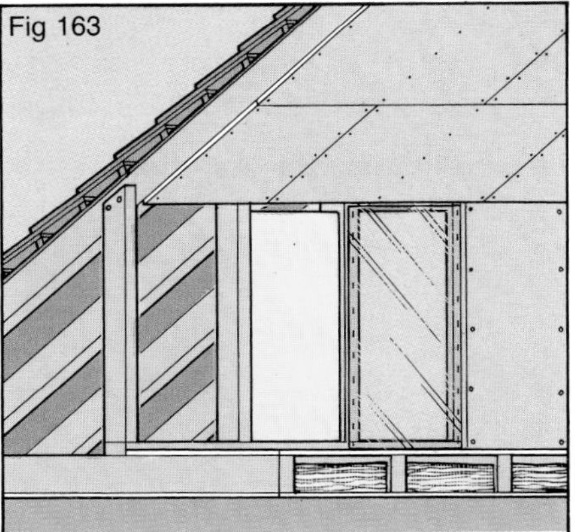

Fig 163 Use the same materials as for the roof slope to insulate the walls of loft rooms. Remove loft insulation from below the room's floor to allow heat to rise into the room from the floor below.

Insulating Flat Roofs

Existing flat roofs are more difficult to insulate than pitched ones because you cannot get at the voids in the structure very easily. You can tackle the job in one of three ways; the one you choose depends largely on the state of the existing roof surface and the ceiling beneath it.

As with pitched roofs, the new Building Regulations lay down strict requirements for both the insulation and ventilation of new flat roofs, but there are a great many flat roofs in existence that have both inadequate insulation and non-existent ventilation. Therefore, if you have a flat roof as part of your property, it will pay you to consider ways of improving its performance in both of these categories.

What to do

You have three main options as far as improving the insulation of an existing flat roof are concerned.

Option 1 involves laying insulation above the existing roof surface and adding a new waterproof surface, creating what the experts call a warm roof deck. You can buy special composite insulation boards for this, complete with a bonded-on layer of roofing felt. These are simply bonded to the existing roof decking and then felted over to complete the new roof surface. It is worth considering this option chiefly if your roof is in need of resurfacing anyway. An alternative is to lay insulation boards over the existing roof surface and then to top it off with lightweight slabs to hold the insulation in place.

Option 2 involves fitting insulation batts or polystyrene slabs between the roof joists. If the fascia boards around the edge of the roof can be removed to allow access to the spaces between the roof joists, then rigid polystyrene boards can be slid into place from the eaves. It is unwise to use batts in this situation, since there will be no vapour barrier above the ceiling and moisture vapour could rise into the ceiling void, condense and saturate the insulation. If the fascias cannot be removed, the existing ceiling below the flat roof will have to be taken down so that insulation can be placed

What you need:
Option 1
- composite insulated pre-felted roofing panels
- bituminous adhesive
- roofing felt
- flashings *or*
- insulation boards
- bituminous adhesive
- lightweight slabs
- sundry tools

Options 2 and 3
- polystyrene insulation boards *or*
- insulation blanket *or*
- batts
- polythene vapour barrier
- plasterboard
- battens
- fixing nails
- sundry tools

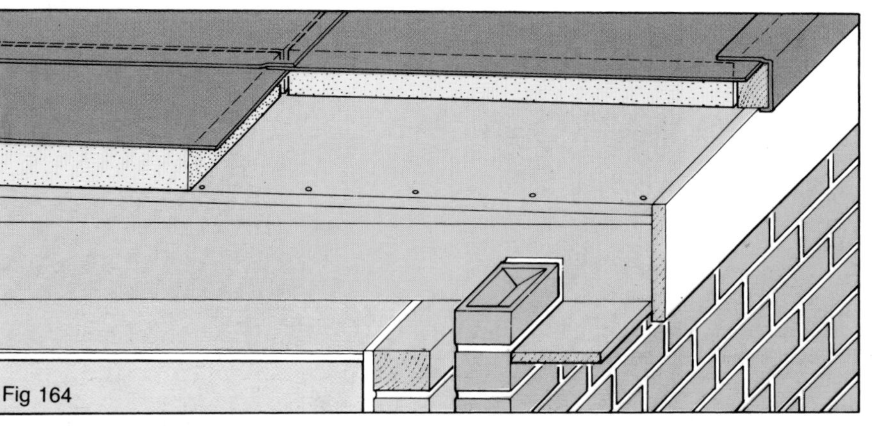

Fig 164

Fig 164 *Option 1* If your roof surface is in poor condition, adding insulation above the roof decking will enable you to provide a new roof surface too. Use composite pre-felted insulation boards, finished off with an additional layer of roofing felt or

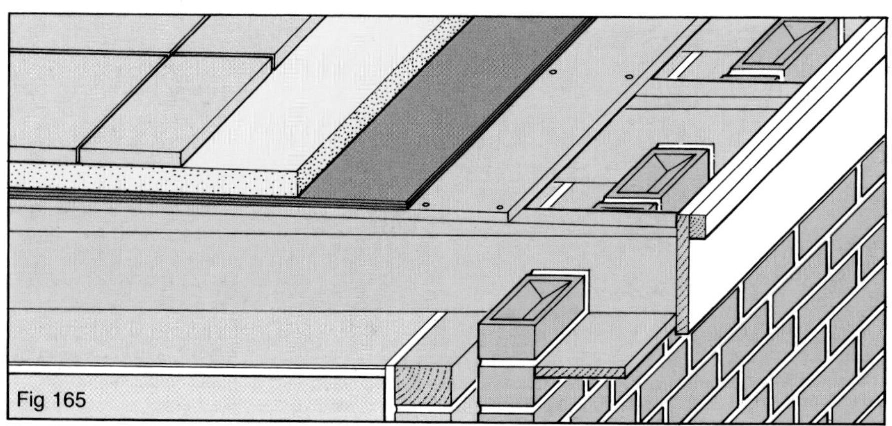

Fig 165

Fig 165 If the roof is in good condition, opt for insulating boards topped with lightweight slabs.

CHECK
- that flat roofs have ventilation openings on opposite sides of the roof equivalent to a slot 25mm (1in) wide running the full length of the eaves

between the joists. A layer of polythene is then stapled to the underside of the joists to act as a vapour barrier. Finally, a new plasterboard ceiling is fixed to the joists. With this method, it is essential to leave a 50mm airspace between the top of the insulation and the underside of the roof deck itself, and to provide ventilation holes on opposite sides of the roof; this guarantees a cross-flow of air through the airspace, preventing condensation from leading to rot in the roof timbers. Choose this option if the roof is in good condition and you do not mind removing the fascia boards or the room's ceiling.

Option 3 is the least disruptive, and involves adding insulation beneath the existing ceiling surface. You can do this in one of two ways. The first is to nail up battens, place insulation blanket, batts or polystyrene slabs between them, add a polythene vapour barrier and then fit a new layer of plasterboard. The second (and the simplest of all) is to fix a layer of insulating plasterboard directly to the existing ceiling.

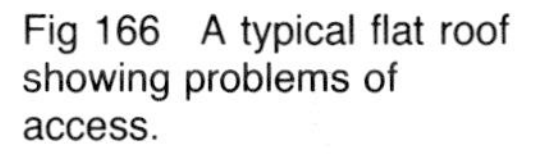

Fig 166 A typical flat roof showing problems of access.

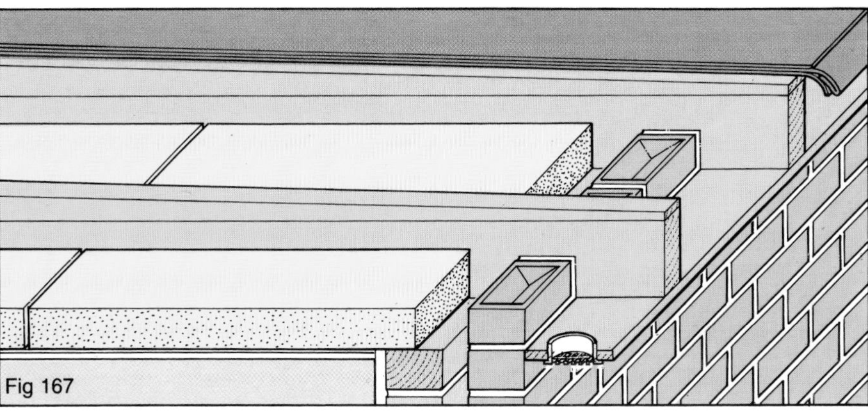

Fig 167 *Option 2* If fascia boards can be removed, slide pre-cut slabs of rigid polystyrene board into the voids between the joists. Alternatively, take down the room ceiling and place insulation between the joists before adding a vapour barrier and a new plasterboard ceiling.

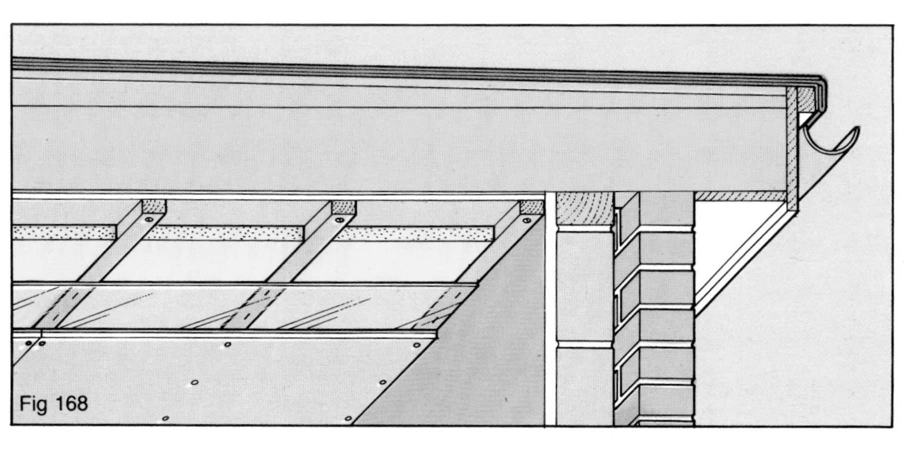

Fig 168 *Option 3* Add insulation below the existing room ceiling, either by placing blankets, batts or rigid polystyrene insulation between battens and covering it with a vapour barrier and new plasterboard, or by putting up a layer of insulating plasterboard.

Insulating Exterior Walls

Your house walls are the most massive part of the whole building and absorb (and lose) a lot of heat, which is why a cold house takes so long to warm up. Solid walls in old buildings are the worst offenders; newer homes with cavity walls perform considerably better, but both will benefit from extra insulation.

You can tackle the insulation of solid walls in two ways: on the inside, or on the outside. Insulating the inside surface is the cheaper of the two alternatives, but does involve fairly major disruption to the rooms concerned (and the loss of a bit of floor space).

What to do

You can use one of two methods. The first involves fixing vertical battens 50mm (2in) square to the wall surfaces, at 400mm (16in) centres. Then insulation blanket, semi-rigid batts or slabs of rigid polystyrene are fitted between the battens, and are covered with a polythene vapour barrier. Finally, plasterboard is fixed to the battens to create a new wall surface.

The second method is somewhat simpler. All you do is fix insulating plasterboard directly to the wall surface. You can use special panel adhesive if the walls are flat, but you will have to put up a framework of slim battens with packing behind hollow spots if they are not.

Whichever method you use will involve removing things like skirtings and architraves, and repositioning light switches and socket outlets, so you must be prepared for the room to be out of use for the duration of the job. There will also be a certain amount of intricate cutting round door and window openings.

External Insulation

If this sounds like an intolerable level of disruption, your only alternative is to consider external insulation. This involves a process similar to that for insulating the inner surface, by placing a layer of insulation next to the wall surface and covering it with some form of exterior waterproof

What you need:

- sawn timber battens
- masonry nails
- carpentry tools
- 50mm thick insulation batts or slabs
- polythene vapour barrier
- staple gun
- plasterboard
- plasterboard nails or
- panel adhesive
- joint tape
- finishing plaster
- radiator foil

CHECK

- that the batten framework is truly vertical, adding packing pieces behind the battens where necessary to cope with hollows in the original plaster surface
- that you overlap successive lengths of polythene vapour barrier generously to minimize the transmission of warm moist air into the insulation

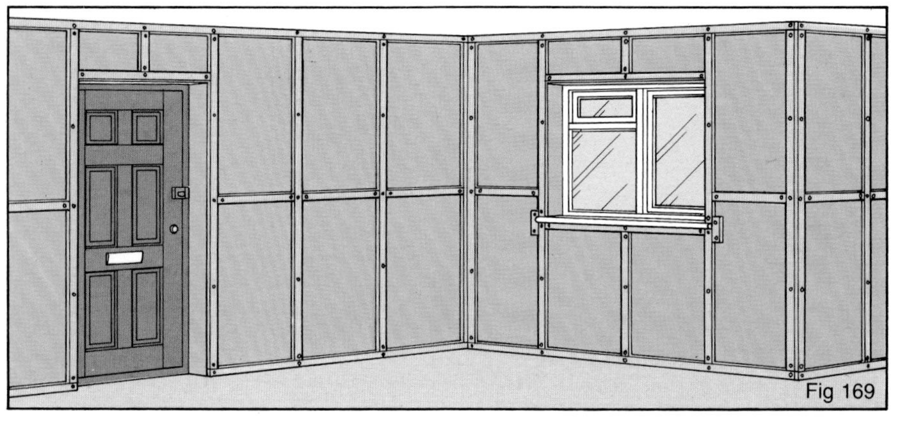

Fig 169 Fix supporting battens to the room's exterior walls. Vertical battens are at 400mm centres, horizontal battens at floor and ceiling level and half-way up the wall. Fit extra battens around window and door reveals.

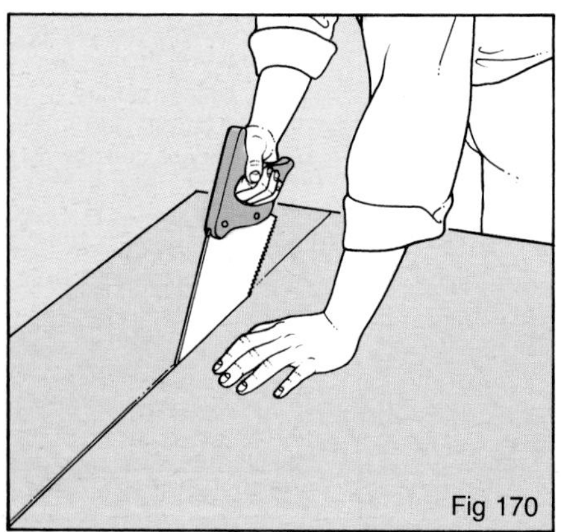

Fig 170 Saw plasterboard to size where necessary using a panel saw, with the sheet well supported on trestles.

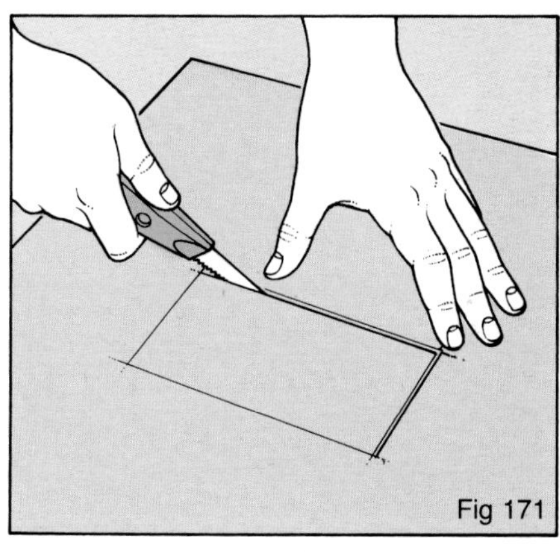

Fig 171 Make small cut-outs for switches and sockets using a pad-saw, after drilling a starter hole in one corner of the marked cut-out.

layer. This may be cladding, tile hanging or a layer of expanded metal mesh, which is then rendered over. You could tackle this yourself, but it is a huge task and is best left to professional installers, although this is likely to be very expensive.

Cavity Walls

You will find cavity walls on almost all houses built since the 1930s, and on a few built before then. The wall consists of two 'leaves' of single masonry with a gap (usually 50mm/2in wide) between them, designed to prevent water penetration. The idea is that any rain passing through the outer leaf will then run harmlessly down its inner face to ground level. Filling the gap with insulating material offers a splendid opportunity to cut heat losses through the wall by up to two-thirds but, since you cannot get at the cavity, it is a job to leave to specialist installers. They will inject insulating material into the cavity through holes drilled in the outer leaf, so that there is minimal disruption to the household.

Fig 172 Semi-rigid insulation batts fit neatly between wall battens, ready for a vapour barrier to be added and the new plasterboard to be fixed.

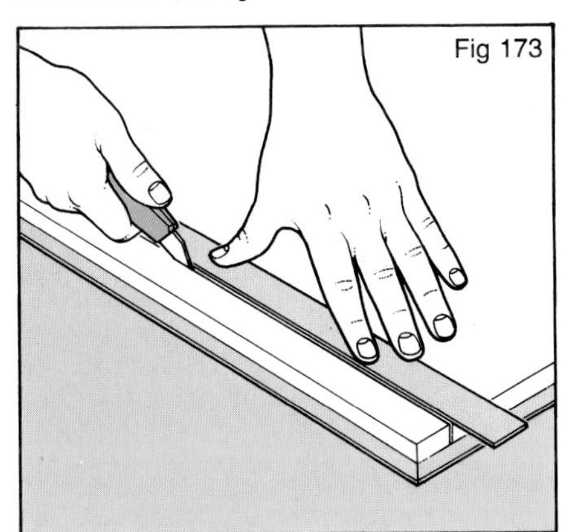

Fig 173 With insulating plasterboard, cut away a strip of the insulation on panels to be used at external corners.

Fig 174 Then offer up the corner-piece so the cut-out fits over the edge of the other board.

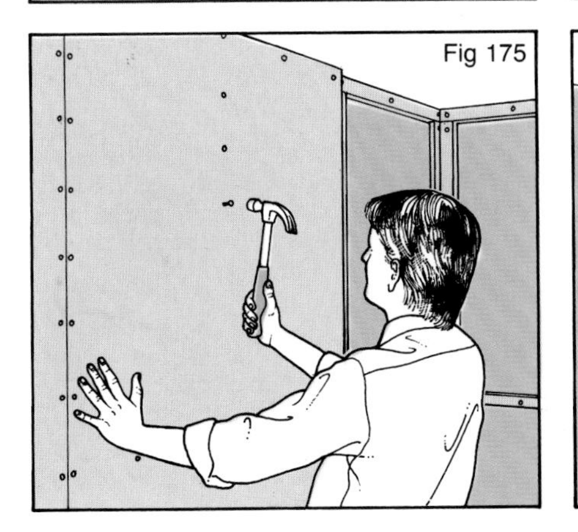

Fig 175 Nail the boards in place to the batten framework using galvanized plasterboard nails at about 250mm centres.

Fig 176 Alternatively, if the walls are flat and true you can bond the panels directly to the surface using panel adhesive.

Insulating Exterior Walls

Several types of materials are used for this, including urea-formaldehyde foam, mineral-wool fibres and specially treated plastic pellets. The only potential source of trouble is that improper installation techniques can lead to damp penetration, since the insulation fills the cavity completely and any voids can provide a cross-over point for water to reach the inner leaf of the wall. For this reason, you should choose a supplier who holds a certificate from the British Board of Agrément, and you should apply to your local authority for permission to have it installed (the Building Inspector will be able to provide you with details of approved installers). Note that cavity wall insulation is *not* suitable for timber-framed homes, which already have insulation built into the wall panels. For more details about cavity wall insulation, *see* page 88.

If you are building from scratch – a home extension, for example – then you can incorporate cavity-wall insulation as the walls rise. You do this using insulation batts held against the inner leaf of the wall by special wall ties, so keeping a narrow gap between the insulation and the outer leaf as a channel for any water that finds its way through the outer skin of the wall to run away harmlessly to ground level.

Insulating Behind Radiators

Radiators work by warming the air in contact with their surfaces. This then rises and sets up convection currents in the room which draw more cool air to the radiator to be heated. They also radiate heat that warms up nearby objects, notably the wall surface behind the radiator. If this is an exterior wall, some of this heat will then be lost through the wall, which is a small but not insignificant heat loss that can easily be reduced.

To do so, either fit panels of special heat-reflecting foil to the wall behind the radiator, or slide semi-rigid panels of foil down behind the radiator so they rest on the wall brackets. Use purpose-made coated foil for this; ordinary kitchen foil will soon tarnish and lose its efficiency at reflecting heat.

TIP
Use a hand-held metal detector to check for buried cable runs in the walls before you start nailing up the battens. Be especially wary of areas vertically above or below all socket and switch positions.

Fig 177 Cavity-wall insulation is injected into the cavity through holes drilled in the outer leaf of the wall. It is a job that must be carried out only by professional installers – *see* page 88 for more details.

Fig 178 It is vital to check that air-bricks are clear after cavity-wall insulation has been installed, to ensure that underfloor voids or rooms containing fuel-burning appliances still have uninterrupted ventilation.

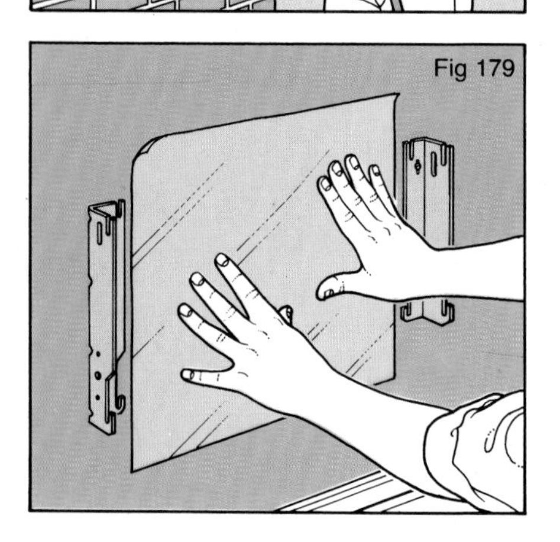

Fig 179 To prevent heat losses occurring through exterior walls on which radiators are mounted, stick special coated reflective foil to the wall surface.

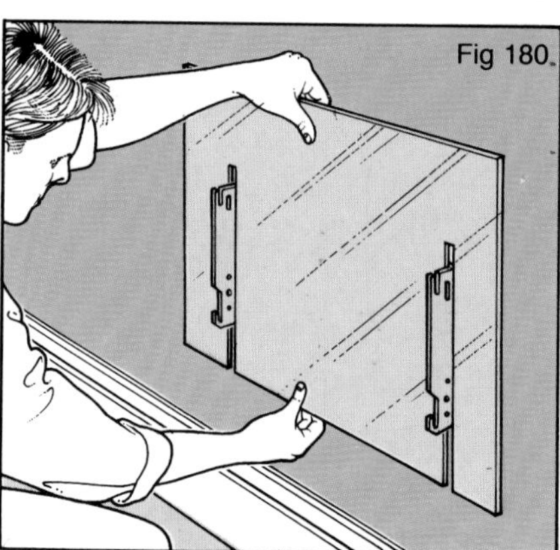

Fig 180 Alternatively, place sheets of semi-rigid coated foil behind the radiator, with cut-outs spaced so that the panel can rest on the radiator brackets.

Insulating Tanks and Cylinders

Most homes in this country have indirect cold-water supplies. This means that taps and WC cisterns are fed from a cold-water storage tank – a useful small reservoir if supplies are interrupted, but a potential nuisance on two fronts. The first risk factor is a leak, which could cause severe damage to furnishings and decorations. The second is a freeze-up, which can occur in cold weather if the tank is in a loft. The risk of a freeze-up is increased greatly if the loft floor is insulated, since the loft will then be deprived of heat rising from the rooms below and, as a consequence, will be much colder in winter.

The solution is to insulate the storage tank, and also the smaller feed-and-expansion tank that tops up a conventional wet central-heating system if this is also sited in the loft. The type of insulation you choose will depend to a certain extent on the shape of your tanks. It should be taken over the top of the tank as well as round the sides. Any loft insulation laid beneath the tank should be removed.

What to do

Start by checking the size and shape of the tank(s). Older homes probably have galvanized metal tanks, and this is a good opportunity to check the condition of the interior. If there are any signs of rust visible, it is advisable to replace the tank with a new plastic one before considering how to insulate it. More modern homes will almost certainly have plastic tanks, either rectangular or circular.

The easiest way of insulating a metal tank is to build a box of insulation around it using rigid polystyrene sheeting at least 25mm (1in) thick. Start by measuring up the tank sides, and cut four panels to size with a sharp knife. Offer up each in turn so you can mark the positions of incoming and outgoing pipes and make cut-outs to accommodate them. Assemble the box, taping the corners, and replace the cut-outs where pipes pass through the insulation. Measure up for the lid, cut the panel and bond a second smaller panel of insulation to it so the assembly is a tight fit.

What you need:
- polystyrene insulation
- measuring tape
- sharp handyman's knife
- tape
- polystyrene adhesive *or*
- insulation blanket
- string
- lid for tank
- gloves *or*
- proprietary jacket for round storage tanks
- tape or string
- proprietary jacket for hot cylinder

CHECK
- that there is no insulation on the loft floor beneath cold-water storage tanks or small feed-and-expansion tanks. Remove any that you find
- that insulation on hot cylinders does not cover an immersion heater thermostat or cylinder stat

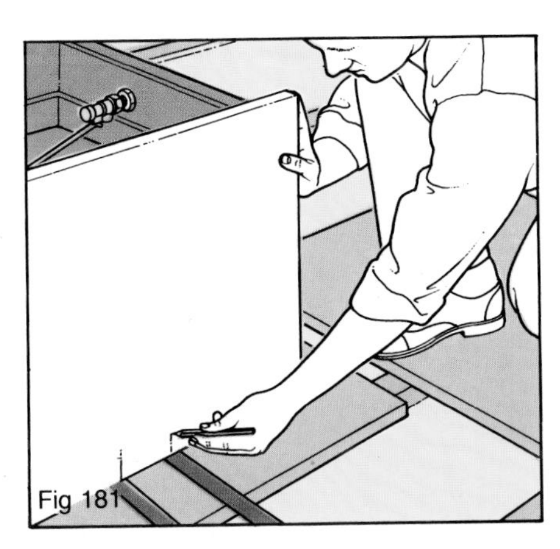

Fig 181

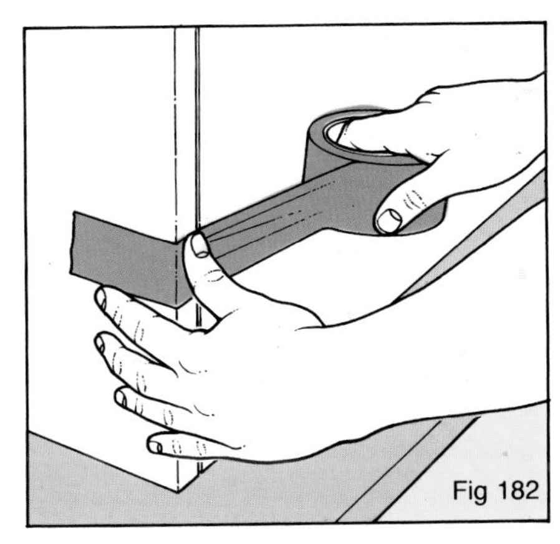

Fig 182

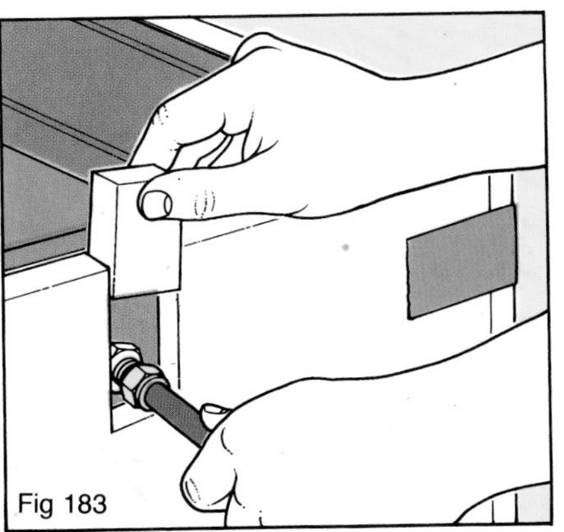

Fig 183

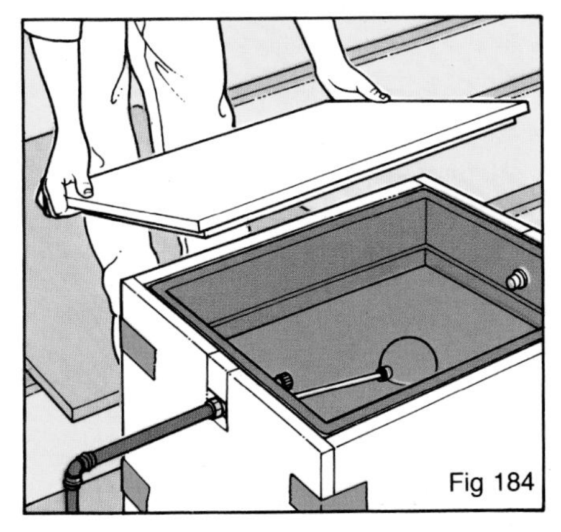

Fig 184

Fig 181 Use polystyrene boards to insulate a rectangular tank. Start by cutting the sides to size, then mark the positions of cut-outs to accommodate incoming or outgoing pipework. Make the cut-outs with a sharp knife, and keep the bits you remove.

Fig 182 Fit the side panels around the tank and secure them in place with adhesive tape.

Fig 183 Cut down and replace the cut-outs at pipe entry points.

Fig 184 Measure and cut a panel for the lid, and stick a smaller piece to its underside to help locate the lid securely on the box.

An alternative for both metal and plastic rectangular tanks is to use the same insulation blanket as on the loft floor. First check that there is no insulation immediately beneath the tank. Then wrap the first length of blanket around the lower part of the tank and secure it in place with lengths of string passed around the tank. Repeat this operation for the second length, again tying it in place. Separate the blanket with your fingers as necessary to fit it tightly around incoming or outgoing pipes.

If the tank has no lid, buy one to fit the tank or make one from a piece of exterior-grade plywood. Do not use chipboard, fibreboard or ordinary plywood, which will be attacked by moisture. Fit the lid in place and then drape lengths of insulation over it, tucking their ends inside the strings holding the insulation around the sides of the tank.

If a vent pipe discharges over the tank, make sure it is long enough to pass below the level of the lid. If it is not you should extend it so it does, making a cut-out in the lid for it to pass through.

Fig 185 (*above*) You can use ordinary insulation materials or ready-made jackets to insulate tanks and cylinders.

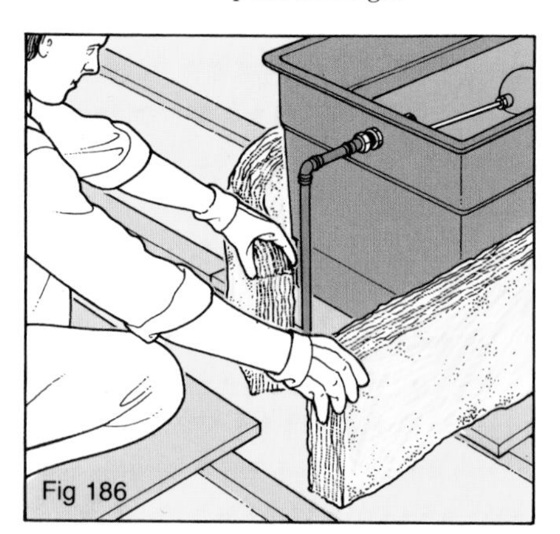

Fig 186 Wrap rectangular tanks in lengths of loft-insulation blanket. Cut the first length to size, wrap it round the tank and secure it in place with string.

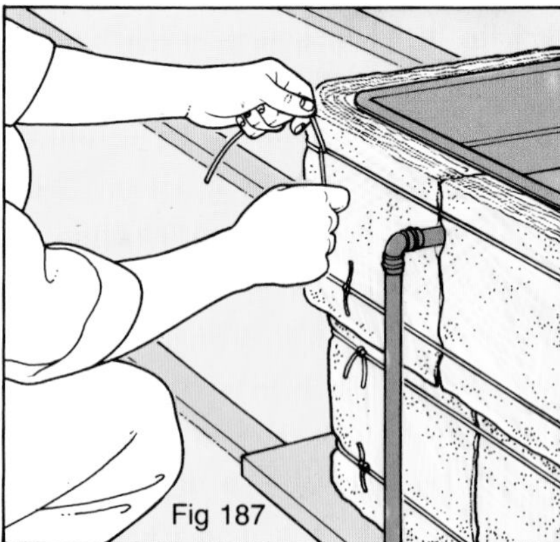

Fig 187 Add a second length, secured in the same way.

Fig 188 If the tank has no lid, buy or make one. It should be a close fit to keep out dust.

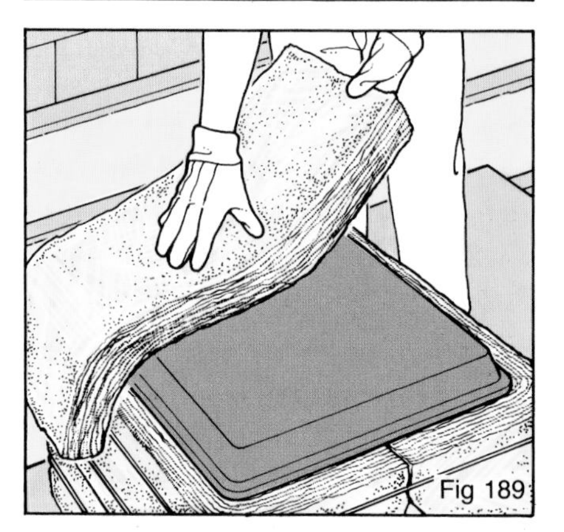

Fig 189 Drape lengths of blanket over the lid and tuck their ends securely into the string loops around the tank sides.

Insulating Tanks and Cylinders

If you have round tanks, the simplest way of insulating them is to buy a ready-made proprietary insulating jacket. These are available to fit all standard tank sizes, and consist of shaped panels of insulation enclosed in a plastic envelope that cover the tank sides, linked to a circular section that fits over the top of the tank. To fit the jacket, simply slip it over the top of the tank, twisting it round so the incoming supply pipe passes between two panels, and then use tape or string to hold the panels in place against the tank walls. Again ensure that any vent pipe present can discharge below the level of the cover.

Insulating hot cylinders

Hot cylinders need insulating to prevent heat loss. Ready-made jackets similar to those used for round storage tanks are available to fit all cylinder sizes. To fit one you simply drape the jacket over the cylinder, tie the top round the vent pipe and then secure the panels around the sides with tape or string.

Fig 190 (*above*) A well-insulated tank will not freeze, even in the coldest weather. Note the vent pipe discharging below the cover.

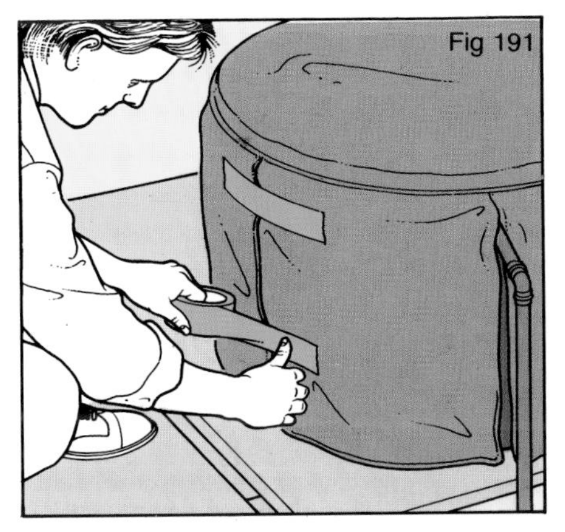

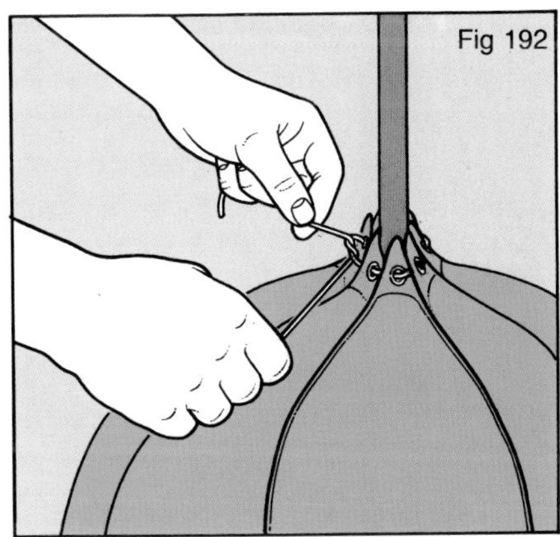

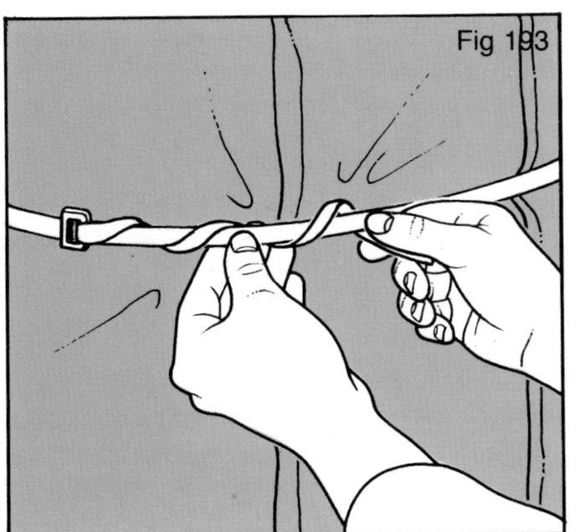

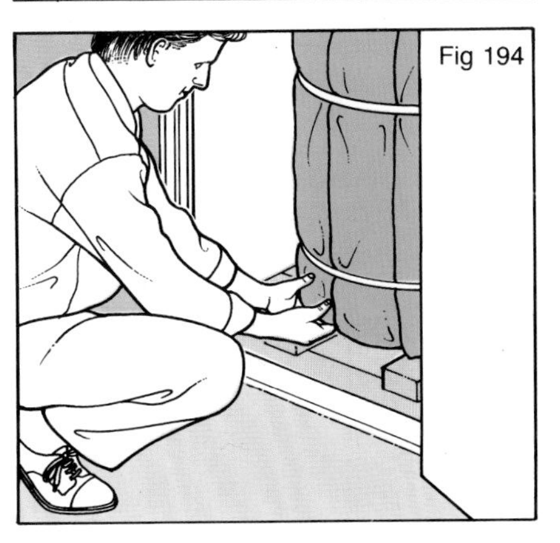

Fig 191 Insulate around storage tanks by fitting an all-in-one insulating jacket. Position it over the tank and then tape the panels together.

Fig 192 Fit a tailor-made jacket over a hot cylinder by tying its neck around the vent pipe.

Fig 193 Then tape the panels together around the cylinder. Make sure you leave an immersion heater thermostat or cylinder stat uncovered.

Fig 194 Finally, check that the panels fit neatly together without gaps. Otherwise cold draughts can easily penetrate these points and cause a freeze-up.

Insulating Pipework

Any pipes carrying water within cold parts of the house need insulating to protect them from freezing in cold weather. This applies especially to pipework in the loft space, below suspended-timber ground floors, and anywhere close to exterior walls, especially if there is an air-brick or ventilator nearby. It is also vital to insulate outside taps and the pipe supplying them, unless you are prepared to isolate and drain down the branch pipework in winter.

In addition, pipes carrying hot water need insulating to cut down on heat losses. This applies to both central heating pipework and pipe runs from the hot cylinder to hot taps.

It is relatively simple to insulate pipes that are accessible, but getting at pipe-runs beneath floors will mean lifting carpets and floorboards, so the job is best left until such time as you carry out major redecoration. You can then tackle it room by room until the entire plumbing and heating system is insulated, but without causing major disruption.

What to do

For long, straight runs of pipework, the easiest material to use is undoubtedly pre-formed foam-pipe insulation. This is available in a variety of lengths – 2m (6ft 6in) is the most common – in sizes to suit 15, 22 and 28mm pipes.

To fit it, you simply open up the pre-cut slit along the length of the insulation and fit it around the pipe. Some types have a positive interlock which you snap closed; cheaper types need taping along the slit to keep it closed. You can take the insulation round gentle curves, taping round it at intervals to keep it in place but, for sharp changes of direction at elbows and tees, it is simpler and neater to mitre the ends of the insulation so that a neat butt joint can be made. Use a mitre box and a sharp kitchen knife to cut the foam, and tape the joints after assembling them.

You can insulate fittings such as stop-taps and gate-valves with the foam by cutting small pieces and taping them round the body of the fitting.

What you need:

- foam insulation to match pipe sizes
- pipe bandage
- adhesive tape
- string or wire
- mitre block and sharp knife for cutting foam
- scissors for cutting pipe bandage
- expanding filler foam
- trimming knife

CHECK

- that joints and overlaps between lengths of insulation are taped so they stay closed, and that insulation runs right up to storage tanks. Otherwise draughts can easily penetrate these points and cause a freeze-up

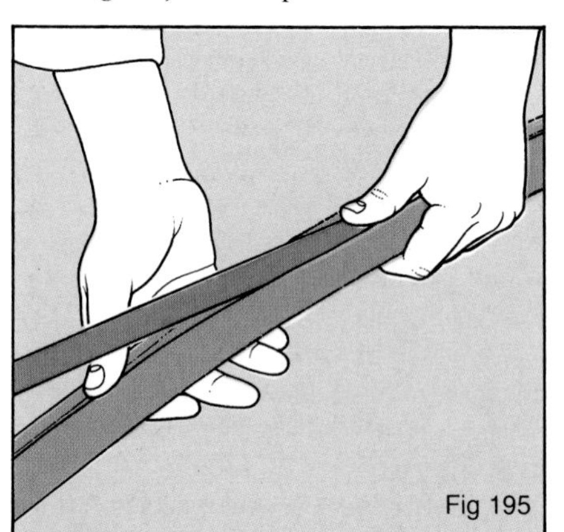

Fig 195

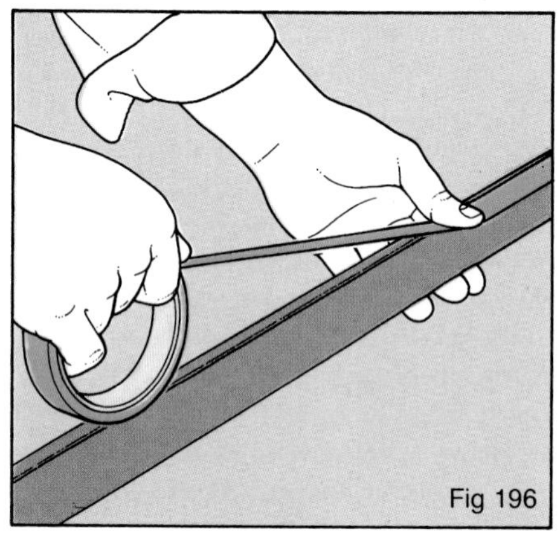

Fig 196

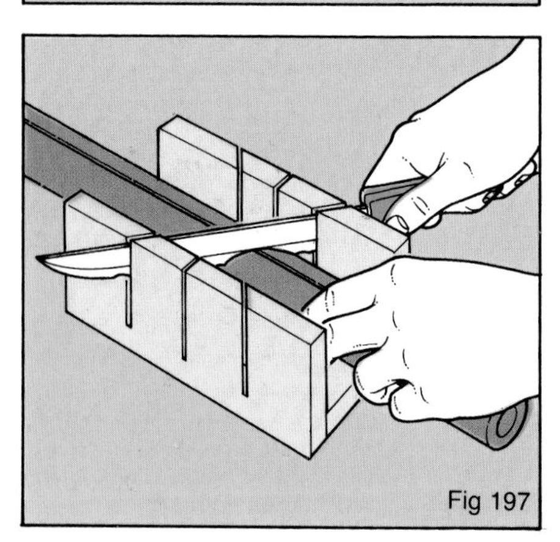

Fig 197

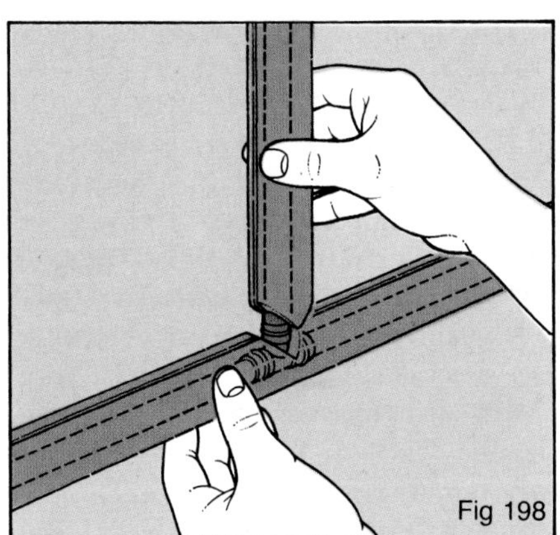

Fig 198

Fig 195 To fit foam pipe insulation, open up the pre-cut slit along its length and fit it around the pipe. Butt-joint successive lengths.

Fig 196 With some types, you press the sides of the slit closed to lock them together. With others it is best to tape along the slit. Tape at butt joints and where the insulation goes around shallow bends as well.

Fig 197 To form neat joints at elbows, use a sharp knife and a mitre block to make 45° cuts.

Fig 198 At tees, cut a V out of the through pipe, and shape the end of the insulation on the branch pipe to fit the V. Again tape the joint.

Insulating Pipework

An alternative to insulation sleeving is pipe bandage, and this is useful for insulating pipe runs with lots of bends and fittings, where it would be very fiddly to make a neat job using foam types. The cheapest type is a coarse mineral wool bandage, while more expensive types have a plastic envelope around them.

Both are wound round the pipe run in a spiral. Start by tying or taping the bandage on the pipe at one end of the run, then wrap it around the pipe so each turn overlaps the previous one. Allow a generous overlap when you join successive lengths, tying the new length in place to stop the previous one from unravelling. Form a figure of eight around fittings such as stop-taps, and finish off the end of the run by again tying or taping the end of the bandage in place.

Expanding filler foam is useful for insulating otherwise inaccessible pipes, especially where they pass through walls. Simply squirt it in, leave it to expand and harden, then trim off any excess. You can use it on outside taps and pipes too.

Fig 199 (*above*) Expanding filler foam is ideal for insulating outside taps and pipework. It is waterproof and can be shaped neatly with a sharp knife once it has set hard.

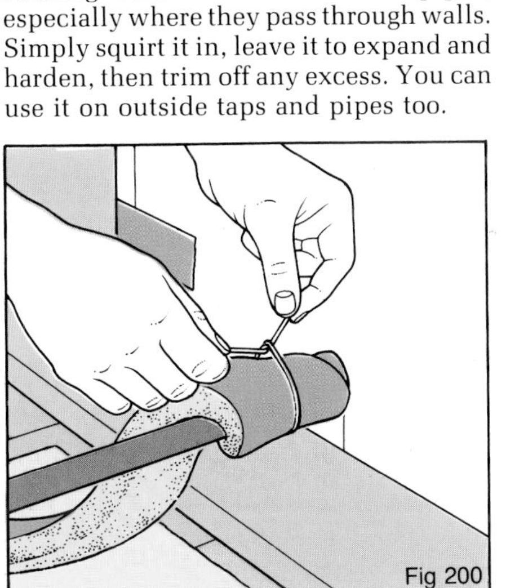

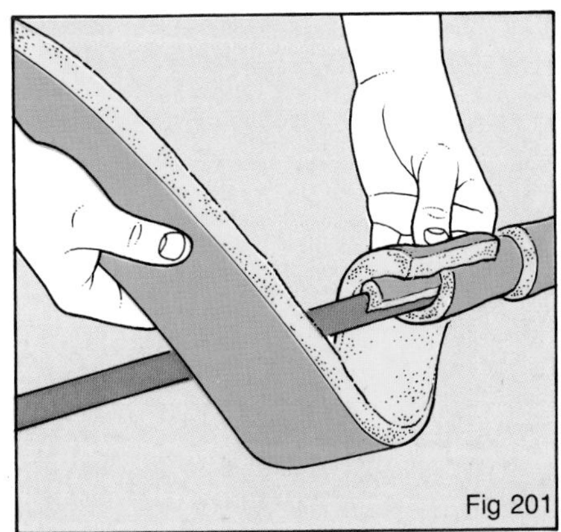

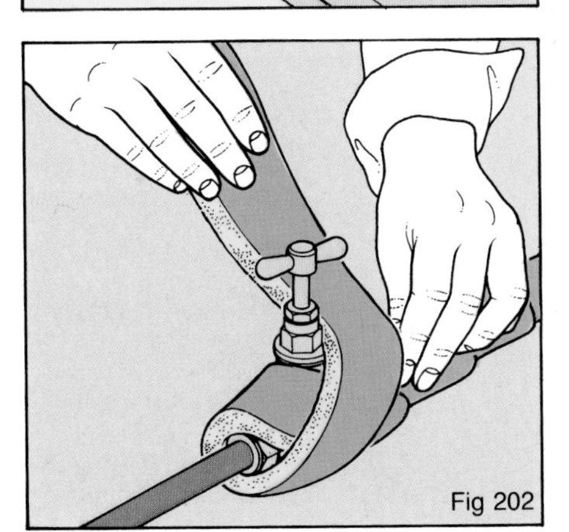

Fig 200 With pipe bandage, tie one end securely to the pipe with string.

Fig 201 Wrap the bandage around the pipe in a spiral. Allow a generous overlap when joining in a new length.

Fig 202 At fittings such as stop-taps, form a figure of eight around the body before carrying on along the pipe.

Fig 203 Use filler foam to insulate pipes in inaccessible places, such as where they pass through walls.

Insulating Floors

Few people think of ground floors when considering insulation, yet a surprisingly large amount of heat can be lost through both solid concrete and suspended timber floors. New homes now have to satisfy the latest Building Regulations requirements for rate of heat loss through ground floors. This is the first time there has been any statutory requirement concerning heat loss through this part of the house structure. In older homes it is possible to introduce floor insulation, but in most circumstances it can be a fairly disruptive job to carry out. You can tackle both suspended and solid floors, but insulating the latter will cause the most upheaval.

What to do

With **concrete floors**, adding insulation means raising the floor level significantly unless you are prepared to dig up the existing floor and re-lay it to incorporate an insulation layer. Adding insulation to an existing floor involves laying 50mm (2in) thick rigid polystyrene boards over the concrete, and then putting down sheets of flooring-grade chipboard on top, making a sandwich which will raise the floor level by about 70mm (2¾in). This will obviously necessitate stripping the room completely, repositioning skirting boards (which will anchor the new floor surface in position once they are replaced), trimming door bottoms and architraves, and forming a small step at the door threshold unless adjacent rooms are also tackled in the same way. You may also have to resite socket outlets at skirting-board level.

If you have a void under a **suspended timber floor** deep enough to crawl into, it is easy to fix insulation between the joists under the floorboards. However, if the void is inaccessible, you have no real choice but to lift the floorboards so you can lay insulation beneath them. If you plan to re-lay or replace the boards anyway, or you are having damp or woodworm treatment carried out that involves lifting the boards, you can take the opportunity to include insulation as part of the overall project. You have several options to choose from.

What you need:

For solid floors
- polystyrene insulation boards
- flooring-grade chipboard
- new skirtings

For suspended floors
- polystyrene insulation board
- supporting battens
- fixing nails *or*
- paper-wrapped glass fibre blanket
- staple gun *or*
- insulation blanket *or* semi-rigid batts
- garden netting
- staple gun
- tools for lifting and replacing floorboards
- saw for cutting boards

TIP

If you are using polystyrene insulation, clip cables beneath the floor to the joist sides clear of the insulation, which can damage the cable sheath if it comes into contact with it.

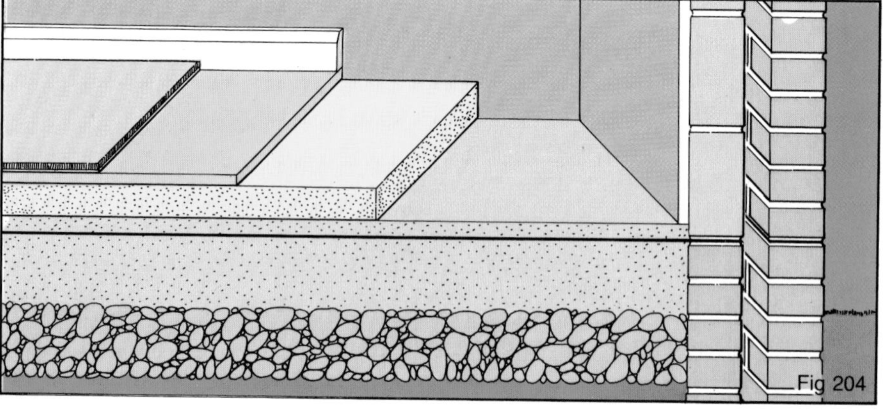

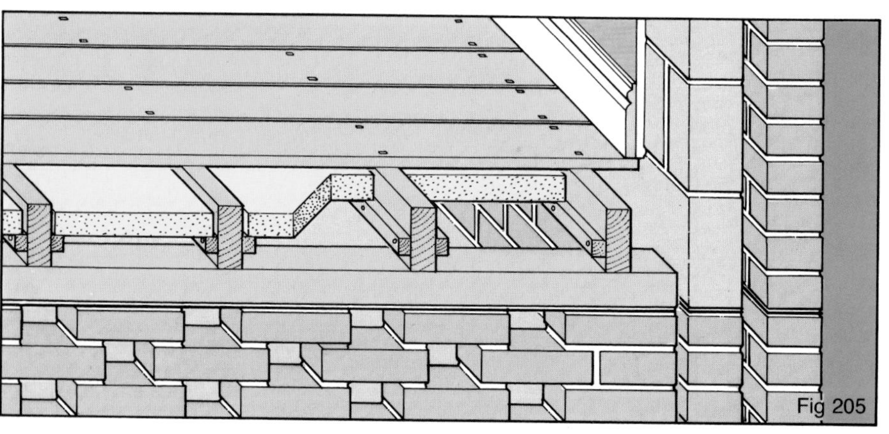

Fig 204 Insulate solid floors by laying expanded polystyrene insulation boards over the screed, and topping them with a new floor surface of flooring-grade chipboard.

Fig 205 Fit slabs of polystyrene insulation board between the joists, supported on slim battens.

Option 1 is to cut strips of rigid polystyrene insulation to match the joist spacing, and to rest them on battens nailed to the sides of the joists. Use boards 50mm thick, and cut them so they are a close fit between the joists. Butt-joint adjacent pieces of board closely together for maximum efficiency.

Option 2 is to staple blanket insulation enclosed in building paper, which is also used to insulate roof slopes, to the joists. This material has paper flanges running along each edge of the roll, making it easy to fix in place using a staple-gun.

Option 3 is to suspend lengths of loft-insulation blanket or semi-rigid batts between the joists using garden netting stapled to the joists to support it. After lifting the floorboards, drape the netting over the joists and staple it to their top edges to leave a slight 'hammock' between adjacent joists. Then lay the insulation in place in the hammocks.

You can also use any of these methods to insulate rooms above ground-floor cold spaces, such as integral garages.

Fig 206 (*above*) Lay garden netting over the joists, stapling it to their top edges, and then place insulation blanket or semi-rigid insulation batts in the 'hammocks'.

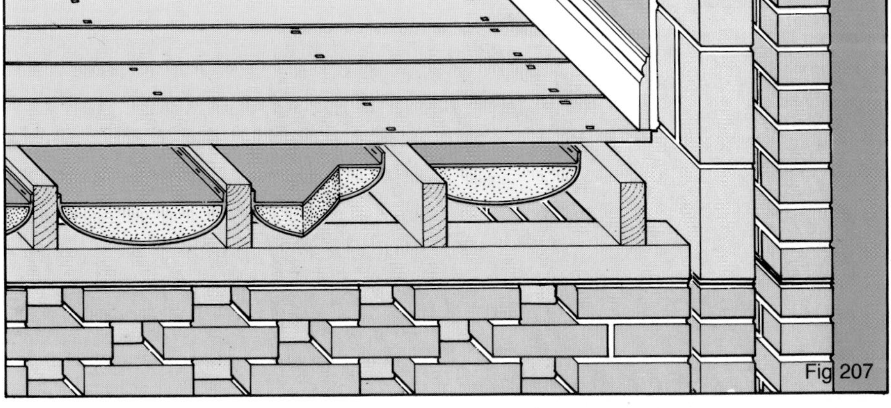

Fig 207 Suspend paper-wrapped insulation blanket between the joists by stapling through the flanges at each edge of the roll.

Using Insulating Underlay

If you do not want to have to strip rooms to lay underfloor insulation, you can still cut down heat losses to a certain extent by laying special foil-faced reflective foam underlay beneath floor-coverings, such as carpet. This will reduce heat loss into solid floors or into the voids beneath suspended timber ones, and will cure condensation, which can rot certain types of floor-covering on cold solid floors.

Lay it just like a normal underlay, taking it right up to the grip strips around the room's edge and taping the joints between adjacent lengths.

Fig 208 (*left*) If floor insulation cannot be installed, put reflective foam underlay beneath carpets to cut heat losses.

Efficient draught-proofing is a vital part of your home's insulation. It will cut down heat losses quite noticeably, does not cost very much to install and will also significantly improve comfort levels around the house. There is just one word of warning: ventilation. Do not overdo the draught-proofing or you might start suffering from condensation and interfering with the safe operation of boilers and other heating appliances.

It goes without saying that badly fitting windows and doors are the cause of the worst draughts. If you have a 3mm (⅛in) wide gap around your front door, that is an area equivalent to having a hole 125mm (5in) square in the front wall, and it is not hard to imagine what a draught that would let in. So do not underestimate the importance of effective draught-stripping for every exterior door, including the thresholds.

Much the same applies to windows, and here old-style wooden sliding sashes are the worst offenders because, by their very nature, they cannot fit too tightly in their frames, or they will be difficult to slide up and down, so they are bound to be draughty. However, casement windows can be almost as bad, especially if the hinged casements and top lights have warped and no longer sit squarely in their rebates. Once again, draught-stripping could block off a lot of incoming cold air and help to reduce unnecessary heat wastage.

What to do: Exterior Doors

You can draught-proof around the sides and across the top of exterior doors using self-adhesive foams or V-strips, in the same way as for hinged wooden windows (see pages 77–78 for more details). However, you will get better results with a wiper or brush-seal type in a metal or plastic holder. Simply pin the strips around the door frame so the wiper or brush seal is compressed when the door is closed. Use the same types on outward-opening doors, fitting the holder to the inside of the frame instead of the outside. Use brush-type draught excluders for sliding doors.

What you need:

For door thresholds
- door bottom excluder or
- threshold excluder or
- two-part excluder with weatherbar

For door frames
- self-adhesive foam strip excluder or
- V-strip excluder or
- wiper or brush seal excluder

For door faces
- letter-box excluder
- escutcheon plate
- scissors
- junior hacksaw
- hammer
- screwdriver

CHECK
- that the seal you fit is well compressed when the door is closed
- that fixing pins on brush types are punched fully home so they cannot snag on hands or clothing

TIP

If you are fitting self-adhesive excluders, clean the door frames down thoroughly with white spirit to ensure good adhesion of the excluder to the woodwork.

Fig 209 (*left*) Give all exterior doors full draught-stripping to the frame and threshold, plus an excluder on the inside of the letter-box opening.

Draught-Proofing Doors

For draught-proofing door thresholds, you have several choices, and which works best will depend on the design of your threshold. For doors opening over a floor surface level with the sill, a spring-loaded brush or wiper seal that lifts up when you open the door and drops down when you close it is the best bet. Where the floor level is slightly below the sill, you can fit a simple non-rising type. Both are cut to length and then fixed to the bottom of the inner door face with screws or nails. With spring-loaded types a small striking plate is fixed to the frame to lift the seal as the door is opened.

An alternative to these excluders is a threshold strip, which fits across the sill and contains a flexible seal that presses against the underside of the door when it is closed. You may need to remove the door and plane it down slightly to get a good seal. However, you can get round this by fitting a two-piece excluder; these have a threshold strip and a separate weatherboard which fits across the external face of the door bottom and forms a water- and draught-tight seal when the door is closed.

Fig 210 (*above*) Brush-type excluders are the most durable types for use on exterior door frames.

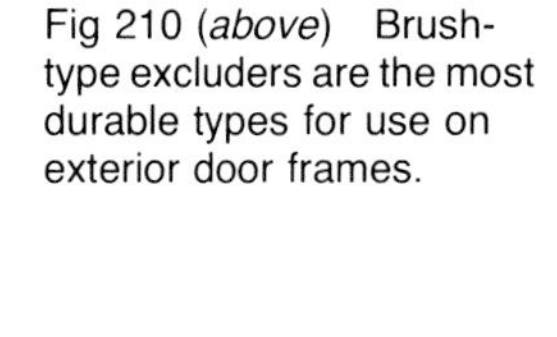

Fig 211

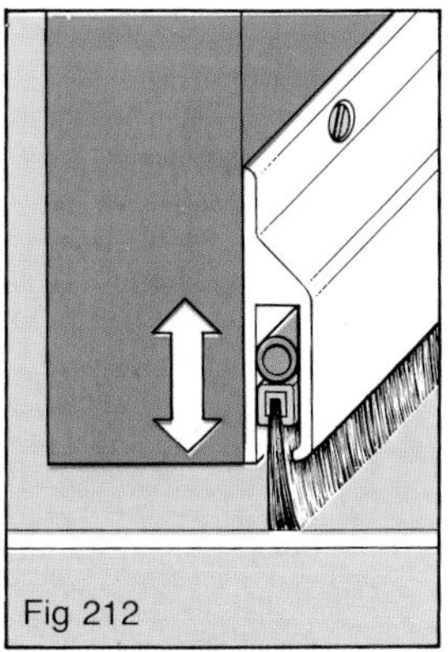

Fig 212

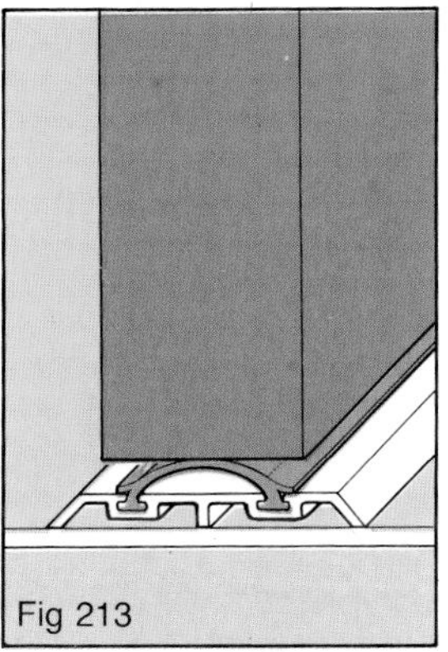

Fig 213

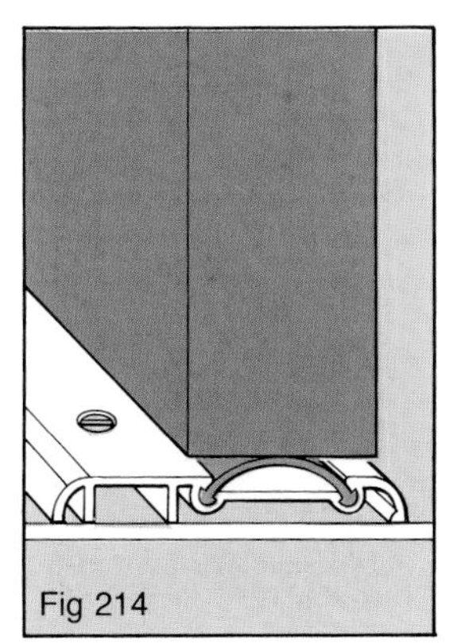

Fig 214

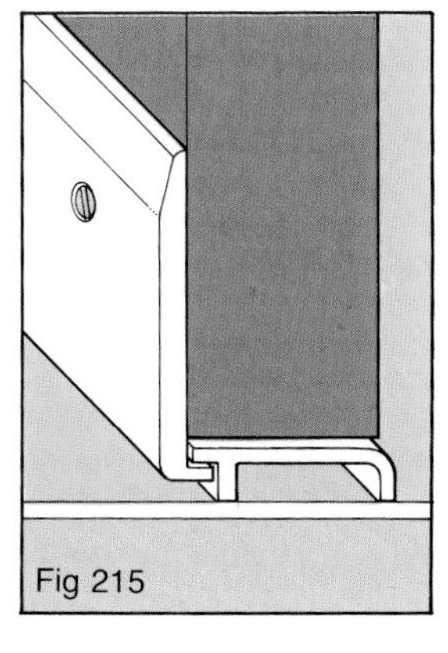

Fig 215

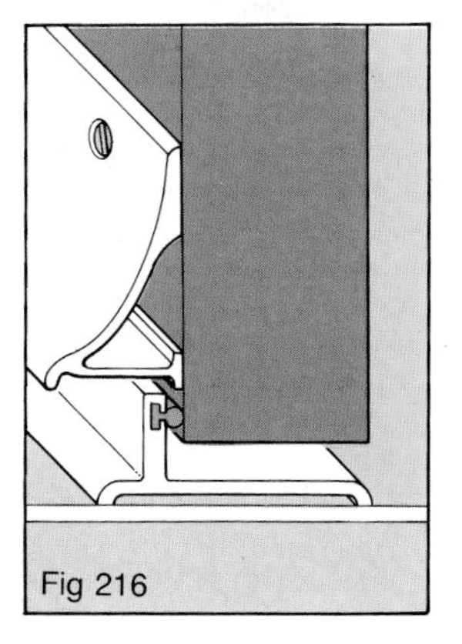

Fig 216

Fig 211 The simplest door-bottom excluder is a brush seal held in a rigid plastic, metal or wooden moulding.

Fig 212 More sophisticated types have seals which lift up as the door is opened.

Fig 213–14 Threshold strips include a flexible seal.

Fig 215–16 Two-part excluders provide additional weatherproofing for doors in exposed locations.

Lastly, if the bottom of your door is rebated and closes on to a weather-bar set in the sill, fit a compression-type excluder along the face of the rebate so it seals the gap between door and weather-bar. This is better than sticking the seal to the inward face of the weather-bar, where the passage of feet will scuff and damage it in time.

Letter-Boxes and Keyholes

The last areas needing attention are the letter-box opening and any keyholes for mortise locks, which pass right through the door and can admit a small but keen draught in cold weather.

For letter-boxes, you can fit a hinged interior flap, a brush-type excluder or one that combines both functions; the brush types have the advantage of keeping out the worst of the draughts even if your newspaper or post is pushed only partway through the opening.

For keyholes, simply fit an escutcheon plate with a cover. These are available in styles to match other door furniture.

Fig 217 (*above*) Tackle outward-opening doors in the same way as hinged casement windows.

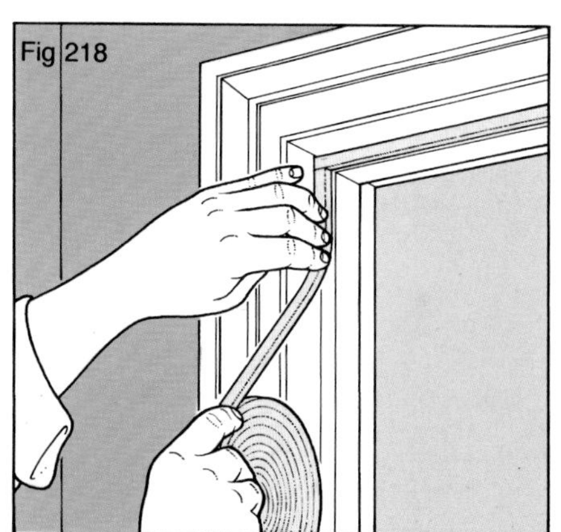

Fig 218 Press self-adhesive foam excluders firmly in place all round the frame rebate.

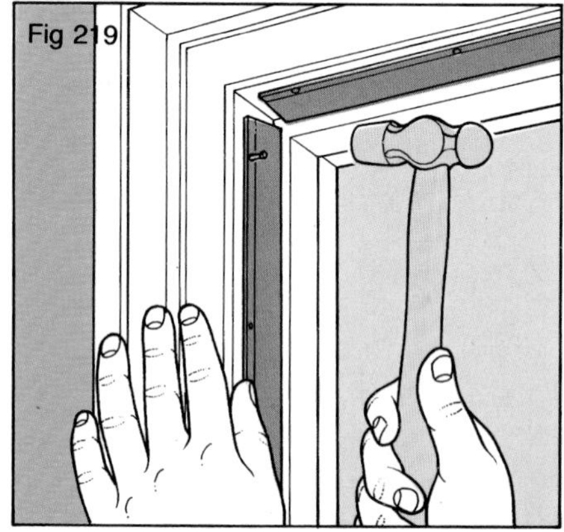

Fig 219 Pin metal V-strips to the frame with the flexible edge nearest the door stop.

Fig 220 Nail brush-strip excluders to the frame so the brush seal will be compressed when the door closes.

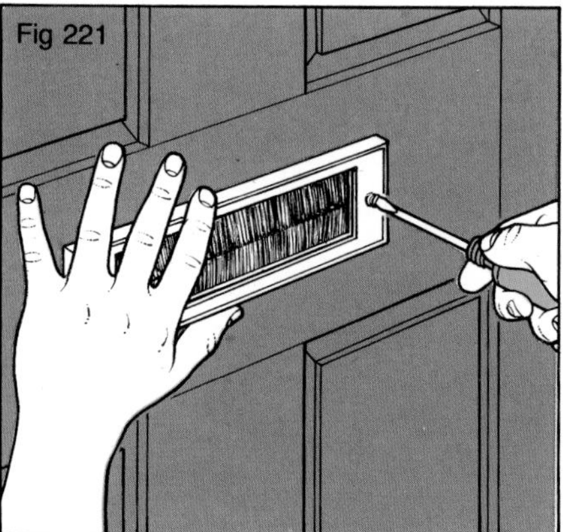

Fig 221 Draught-proof letter-boxes by fitting an excluder over the opening on the inner face of the door.

What to do: Hinged Windows

For hinged wooden windows, the cheapest and most effective type of draught excluder to choose is one of the self-adhesive compression types. You simply stick this to the face of the rebate into which the window closes (except down the hinged edge, where it should be stuck to the side of the rebate so it is compressed by the edge of the closing window). These products have come a long way from the crumbly foams of twenty years ago; most are now made from either PVC foam or from EPDM (ethylene propylene diene monomer), and have profiles variously described as E, P and K because their cross-sections resemble those letters.

The other type of compression excluder available for use around wooden windows is the V-seal, a strip of metal or plastic folded into a V-shape which is pinned or stuck around the rebate so that the raised part of the V is compressed as the window closes. It's a fiddle to fit, the metal types can have sharp edges and the plastic ones may become brittle in time, but they are a durable alternative to self-adhesive foam types.

More expensive are the brush-strip types, similar to those used on door frames. These can be used on any type of window, but are essential on sash windows and horizontally sliding metal windows (see below).

If you have metal windows, you cannot usually use self-adhesive foam types

What you need:
For hinged windows
- self-adhesive PVC or EPDM foam excluder or
- V-seal excluder or
- wiper-seal or brush-seal excluder

For sash windows
- compression seal excluder for horizontal surfaces
- wiper-seal or brush-seal excluder for vertical surfaces
- V-seal excluder for meeting rails

For horizontally sliding windows
- compression seal excluder for vertical surfaces
- brush-seal excluder for horizontal surfaces
- scissors
- junior hacksaw
- hammer

Fig 222 Self-adhesive foam excluder is the cheapest and simplest solution for casement windows.

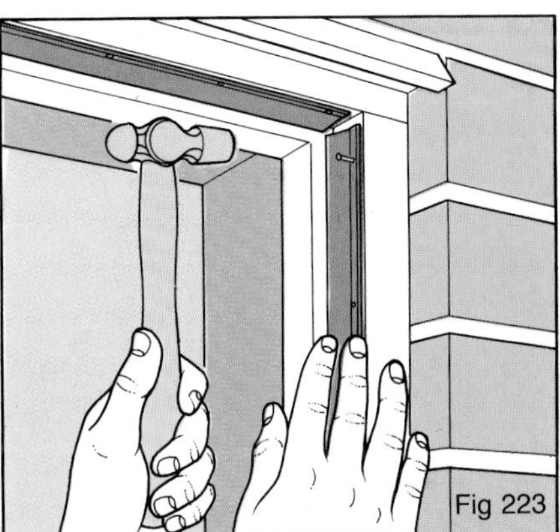

Fig 223 Alternatively, use pin-on or self-adhesive V-seal excluder.

Fig 224 Wiper- or brush-seal excluders can be pinned or stuck in place.

Fig 225 Use silicone sealant for draught-proofing metal windows and warped wooden ones. Pipe it around the frame, then close the window.

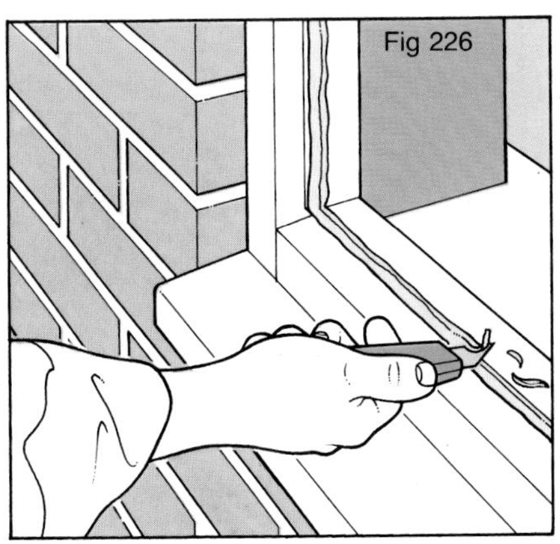

Fig 226 When the sealant has set, trim the excess with a sharp knife.

because the casements close too tightly within the frames. The best type of draught-proofing to use is silicone sealant. You pipe this around the rebate from a caulking gun, with the window open. Then you stick special low-tack tape on the inner face of the hinged casement, and close it to compress the mastic so it accurately fills any gaps. After twenty-four hours, open the window, peel off the tape and trim off any excess sealant.

What to do: Sliding Windows

With vertical sliding sashes, you need three types of excluder to make the sashes effectively draught-proof. First of all, use any of the compression types mentioned under **Hinged windows** to seal the gap where the top and bottom sashes meet the horizontal parts of the frame. Next, fit a flexible brush or wiper seal held in a plastic or metal channel to the inside of the frame so that the seal presses against the inner face of the bottom sash. Use the same type on the outside of the top sash, choosing a wiper seal rather than brushes here since rain may wet the brush pile and make it less efficient at closing gaps. Finally, seal the gap where the two centre rails meet with either V-strip fixed to one sash or brush seals pinned to both. You can help reduce any rattling, and pull the two sashes tightly together when they are closed, by fitting a Brighton pattern screw-up window fastener instead of a conventional fitch-type closer.

For horizontal sliding metal windows, use compression seals at each side of the frame and brush seals at top and bottom.

Fig 227 (*above*) Self-adhesive V-seal excluder is an economical alternative to foam types.

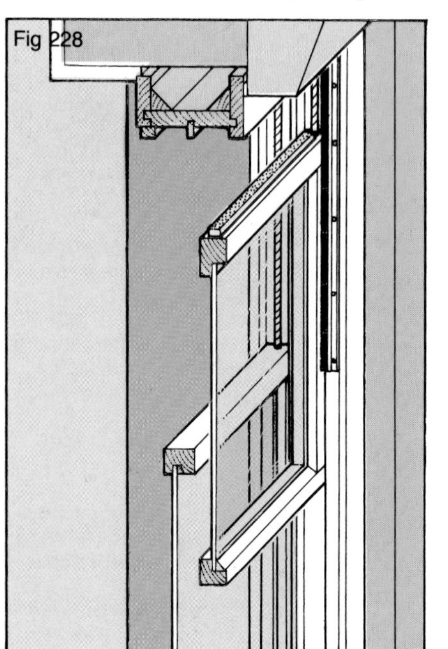

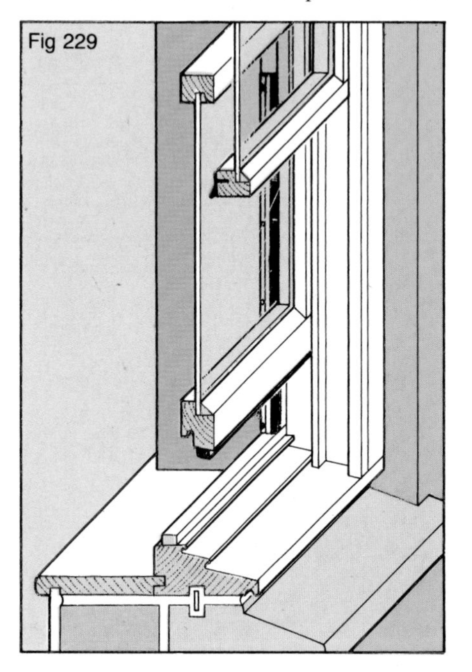

TIP
Seal up windows that are never opened in winter by piping silicone sealant around the frame where it meets the casement or sash. The sealant can simply be peeled off in the spring.

Fig 228 To draught-proof a sash-window, fit foam excluder on the top edge of the upper sash and wiper- or brush-seal excluder to the frame on the outside.

Fig 229 Similarly, fit foam excluder to the bottom edge of the inner sash, and wiper- or brush-seal excluder to the frame on the inside. Fit V-seal excluder to the meeting rail.

Other Cold Spots

Every house has plenty of other places where draughts can get in, and which may need attention. Starting at ground-floor level, you may have noticed dark tramlines across pale fitted carpets laid over suspended timber floors. The cause of this is air blowing up between the floorboards and bringing dust with it. Do not tackle it by blocking up the underfloor air-bricks, because that void needs ventilation or the timbers may develop rot. Instead, lift the carpet and put down paper or hardboard over the existing boards to block off the air flow.

It is also worth tackling gaps between skirting boards and the floor surface. One of the simplest methods is to pin lengths of slim quadrant moulding to the skirting board to close off the gap.

While you are at it, fill holes where pipes or cables pass through floors, especially in cupboards and under stairs. Do the same at first-floor ceiling level, where pipes and cables drop down from the loft.

If rooms have air-bricks set in the walls that are a source of cold draughts, check how essential they are by temporarily taping some polythene over them. If condensation problems do not result, replace the polythene with a plastic hit-and-miss ventilator, which can be closed in cold weather. Do not block air-bricks in a room containing a fuel-burning appliance, though, unless you are making some alternative provisions for its air supply. It is wise to get a professional heating and ventilating engineer to check that your appliances have sufficient ventilation for them to burn safely.

Remember that by draught-proofing all these unofficial sources of fresh air, you might also be introducing a condensation problem. In steamy rooms like kitchens and bathrooms, you should think seriously about installing extractor fans, but for other rooms a little controlled ventilation may be all that is required.

A neat way of providing draught-free ventilation is to fit what are known as trickle ventilators above existing window frames (*see* page 84 for more details).

One last source of draughts is the trapdoor leading to your loft. Assuming that your loft is properly insulated and well ventilated, the loft will be a pretty cold place. You can draught-proof it by using any of the draught excluders mentioned earlier for use around windows.

Double glazing is just what its name suggests: a second layer of glazing material fitted in a window next to the existing glass. The trade calls it secondary glazing. It works by trapping still air between the panes, and it is this air sandwich that improves the insulation performance of the window so dramatically. Many new and recently built homes have their windows fitted with one-piece sealed double glazing units which do the same job, but it is not always possible to fit these into existing frames, so secondary glazing is the only alternative.

The commonest DIY double glazing systems consist of a kit of special mouldings – either in plastic or aluminium – which you use to 'frame' panes of glass bought from your local glass merchant, or panels of clear rigid plastic sheeting (*see* CHECK) which you can buy from DIY stores and cut to size yourself. This framed pane is then mounted on the inside of your existing window frame, either on hinges, or in special tracks that allow the panes to slide open and closed. Some of the cheaper

CHECK
- that you choose the right plastic for secondary double glazing. The range available includes: **semi-rigid PVC sheet** – the cheapest, but is not fully clear and will degrade in sunlight unless stabilized; **polystyrene sheet** – about two-thirds the price of glass, will also degrade in strong sunlight; **acrylic sheet** – better known as Perspex, costs about the same as glass, comes in colours as well as in clear form; **polycarbonate sheet** – the clearest type, also available in colours, but costs about twice as much as glass

Fig 230 Sliding panels.

systems are designed to be fixed to the frame with strong double-sided adhesive tape, magnetic fixings, small turn-button clips or screws but, for safety reasons, types secured with screws should never be used in rooms where the window might be needed as a fire escape route. With all these types, the panels can be left in place all year round, or can be removed in the summer months.

As a very cheap alternative to 'solid' double glazing, there are several systems on the market using then flexible plastic film which is held in place with double-sided tape, and which you put up in the autumn and remove (and throw away) in the spring. This type is particularly suitable for seldom-used windows, and for people with limited budgets or living in rented accommodation.

What to do: Framed Types

With all framed types, be sure to read the instructions supplied with the kit carefully before you start, and check that you have all the kit components.

With **tracked types** start by cutting the track mouldings to length to fit the top, bottom and sides of the window reveal or frame. Then follow the guide-lines given in the instructions about measuring up for the sliding panes, and order the glass (if you are using plastic sheet, you can, of course, cut it yourself).

TIP
1 Clean your windows thoroughly before installing double glazing, especially the thin film or fixed pane types.
2 Fit draught-stripping to the existing window, so that the trapped air between the panes stays still and can do its job as an insulator.
3 Keep the panes closed whenever possible to minimize the occurrence of condensation between the panes.

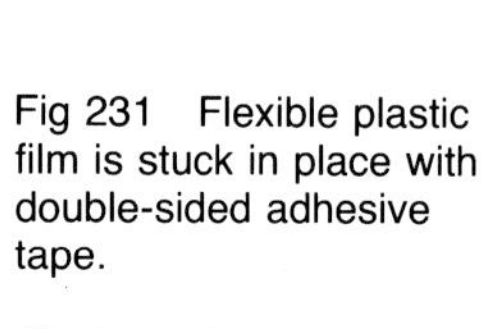

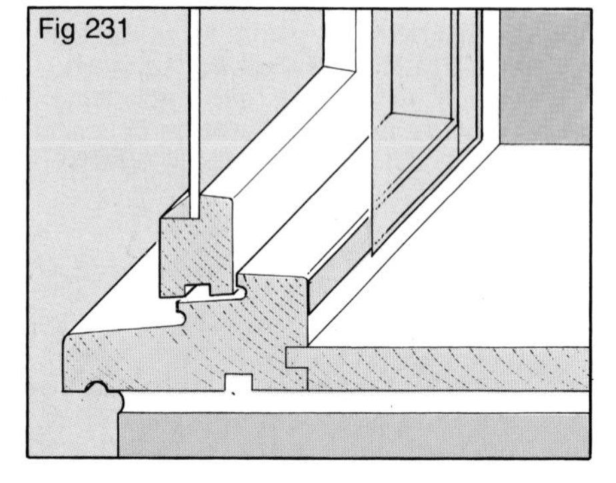

Fig 231 Flexible plastic film is stuck in place with double-sided adhesive tape.

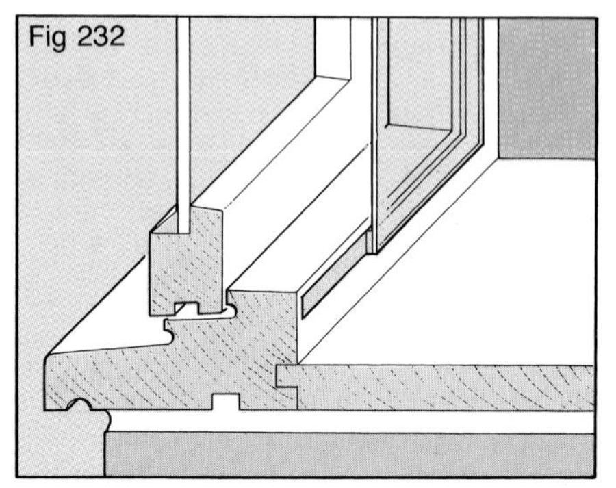

Fig 232 Rigid plastic sheeting can be held in place to the frame with self-adhesive magnetic strips, stuck to both frame and pane.

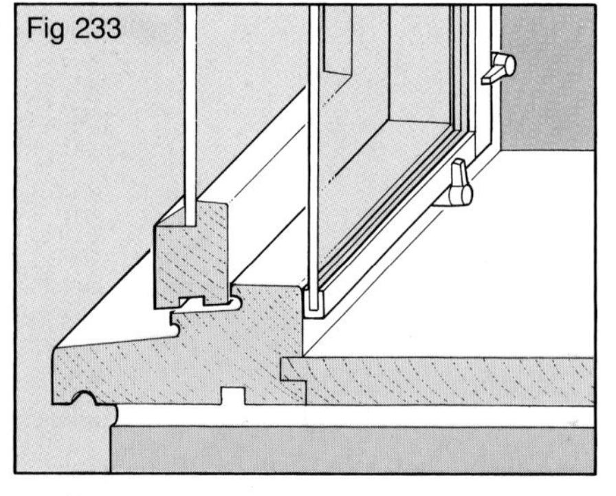

Fig 233 Framed panes can be secured to the window frame with small turn-button clips.

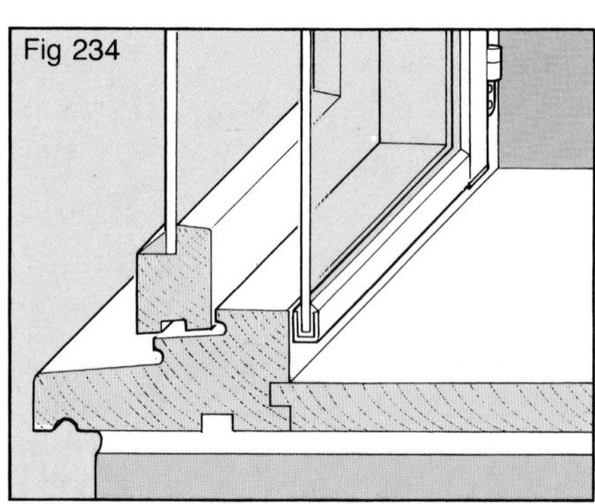

Fig 234 Hinged types can be opened for ventilation or held closed with a clip on the side opposite the hinges.

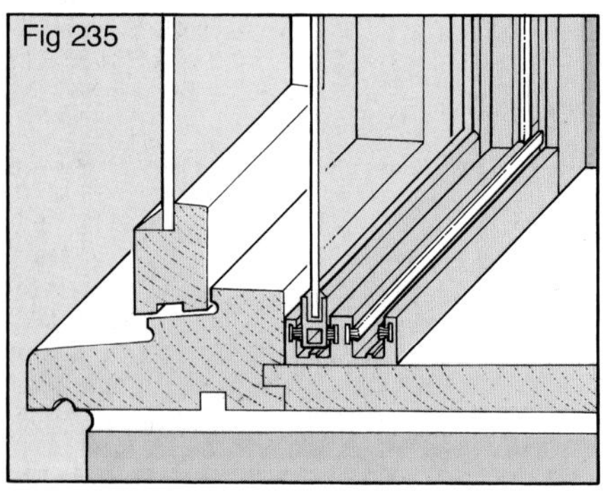

Fig 235 Tracked types consist of two framed panes mounted in top and bottom tracks.

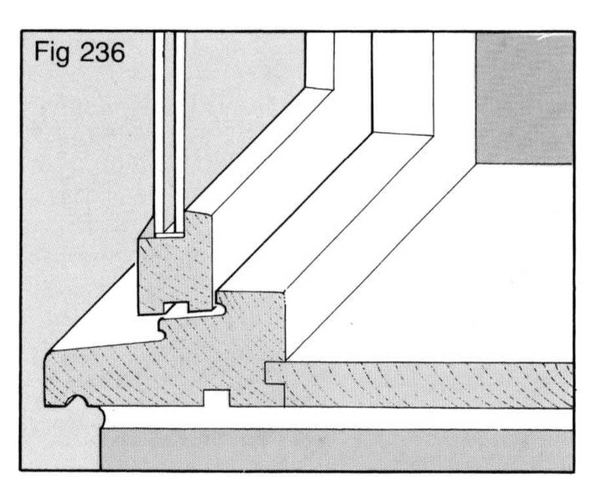

Fig 236 Sealed unit double glazing consists of two panes of glass factory-sealed all round the perimeter. The units fit into deeper-than-usual rebates, and may be retained by putty or beading.

Cut the mouldings that fit around the edges of the pane, and tap them into place after fitting their gaskets to the pane edges. Make sure that the draught seals are correctly placed. Then add the corner-pieces and lift the framed panes into place in their tracks so you can check that they open and close properly. Lubricate the tracks with furniture polish if they don't slide easily.

With **hinged types**, frame the panes as for track systems, and then mount the pivot hinges on the frame so you can hang the pane in place. Again, make sure the draught seals are correctly positioned. Add the catches to the frame at the opposite side to the hinges to keep the pane closed against the frame.

With **fixed types**, again frame the pane as per the instructions and then secure it to the window frame using double-sided tape, magnetic strip, clips or screws, according to type.

What to do: Thin Films

If you choose this disposable type, start by cleaning the window frame thoroughly to remove dirt and grease. Then apply double-sided adhesive tape all around the perimeter, and cut a piece of film slightly larger than the frame size.

Stretch the film across the top of the frame, and press it on to the tape. Then draw it down to the bottom, keeping it as taut and crease-free as possible, and fix it to the tape at the sides and bottom of the frame.

Next, play hot air from a hair-drier over the film to shrink it and pull out any creases. Finally, trim off the waste all round with a sharp knife.

Fig 237 (*above*) Most types of secondary double glazing can be easily removed and stored away during summer months.

Fig 238 Fix thin-film types in place using double-sided tape, then use heat from a hair-drier to remove the creases.

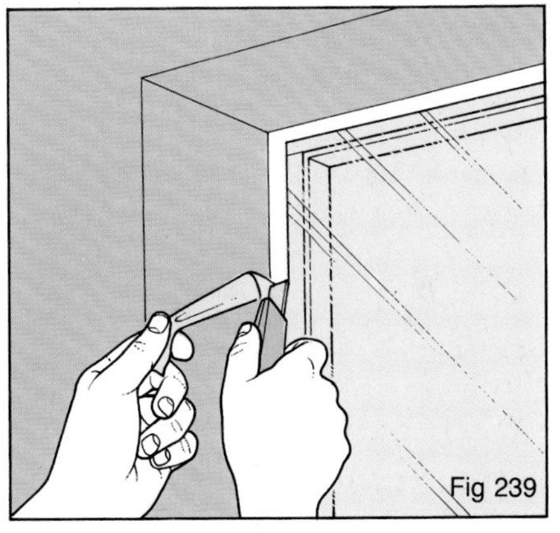

Fig 239 Trim off excess film all round with a sharp knife. Peel the film off in summer.

Noise penetrating through party walls from the house next door is one of the biggest sources of disputes between neighbours, and the problem is worst of all in relatively modern homes of lightweight construction. The latest Building Regulations go some way towards improving matters by specifying minimum masonry densities and by banning practices like the building-in of floor joists into party walls but, in many existing homes, additional sound-proofing is the only way of improving the wall's acoustic performance.

What to do: Party Walls

The most significant improvement can be gained by adding insulation on the noisy side of the wall, so, if your neighbours keep their TV or hi-fi turned up, it is worth approaching them (if you are still on speaking terms) and asking whether they are willing to line their side of the wall with sound-absorbing material such as cork or acoustic tiles. You could even suggest sharing the cost.

If they are not, you will have to tackle the problem from your side. Start by examining the structure of the existing wall, since even the smallest cracks and holes can let sound pass. You are most likely to find these in the loft, where pointing may be of poor quality and there is no plaster to seal the gaps. Repoint the wall if necessary, and add a skim coat of plaster over the whole wall surface.

Next, lift floorboards adjacent to party walls and check for gaps there, especially where floor joists are built into the wall. Pack gaps between the joists and the masonry with mortar, and fill any other cracks or voids you find. Then seal any cracks between walls and ceilings and between skirting boards and floors using non-setting mastic.

If this does not produce any significant improvement, you will have to take further steps. The most cost-effective solution is to erect a sort of stud partition wall on your side of the party wall. To reduce flanking transmission, this must be constructed so that it does not touch the party wall.

This means fitting head and sole plates to the ceiling and floor, and adding 75 × 50mm (3 × 2in) vertical studs between them at 600mm (2ft) intervals so that there is a free air space between the rear face of the framework and the party wall. You then suspend lengths of 100mm (4in)-thick glass-fibre insulation blanket behind the studs, and line the partition with two thicknesses of 13mm (½in) tapered-edge plasterboard. Stagger the joints between the two layers, and tape and fill the joints in the final surface layer to leave a smooth,

What you need:

For party walls
- 75 × 50mm (3 × 2in) sawn timber
- 100mm (4in)-thick insulation blanket
- 13mm (½in)-thick plasterboard
- fixing nails
- joint tape and filler
- non-setting mastic
- carpentry tools

For floors
- insulation blanket
- battens for raft floor
- flooring-grade chipboard
- carpentry tools

For windows
- kit for sliding secondary glazing
- acoustic tiles
- sundry tools

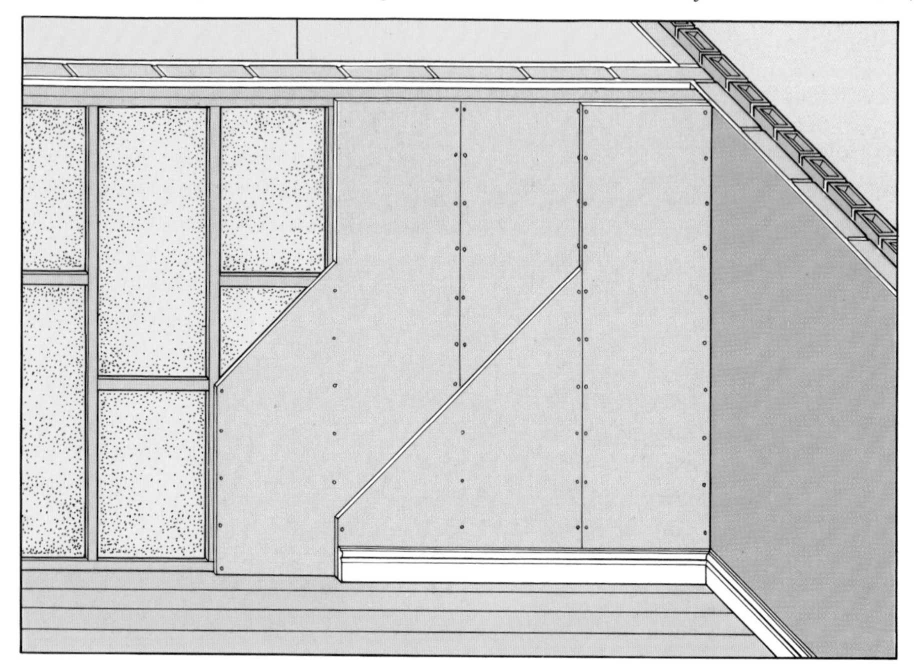

Fig 240 To insulate a party wall, erect a stud partition wall in front of it, with insulation blanket behind it and two layers of plasterboard on the framework.

seamless finish. Finish off by adding new skirting boards to match the existing ones, and seal any slight gap between the new wall surface and the ceiling with mastic.

Building a wall of this sort will mean fairly major disruption and some loss of floor space but it will cause a significant reduction in noise transmission, and is relatively inexpensive to construct.

What to do: Floors

Noise travelling between floors inside a house is rarely a problem, but can rapidly become one in buildings that have been converted into flats.

The best way to sound-proof an existing floor is to create a floating floor. This involves constructing a new floor surface separated from the existing floor with a continuous layer of glass fibre or mineral wool quilt. The new floor can be a raft or a platform. The former is constructed by lifting the existing floorboards and draping the quilt over the joists. You then locate 50mm (2in) sq battens over the tops of the joists and fit a new floor surface consisting of two layers of 19mm (¾in) flooring-grade chipboard. Turn the edges of the quilt up so they are sandwiched between the walls and the edges of the floor surface, and conceal them with new skirtings.

The platform-type floor is easier to lay, and raises the floor level in the room by less than the raft type. Here the quilt is laid on top of the existing floorboards, and a new deck of 19mm chipboard or plywood is laid directly over it. Again, the quilt is turned up at the room perimeter to insulate the new floor from the rest of the structure.

What to do: Windows

Windows are a particularly weak link as far as insulation against noise from outside the house is concerned – particularly from traffic and aircraft. The solution is to install acoustic double glazing. Try to choose hinged or sliding panels which can be opened for ventilation; you can use the same types as for thermal double glazing, with the very important difference that the new panes should be set 150–200mm (6–8in) away from the existing window, which should be efficiently draught-proofed (and kept closed). In addition, the sides, top and bottom of the window reveal should be lined with acoustic tiles.

TIP

If you live near an airport or a major road, you may be able to obtain compensation or even a grant towards the cost of sound insulation. Contact your local authority or a Citizens' Advice Bureau for details of schemes operating in your area.

If your problems seem to defy DIY solutions, call in a specialist acoustic consultant for advice. Look up local firms in the Yellow Pages under the heading 'Acoustic engineers', or else contact a body called the Association of Noise Consultants (Spectrum House, 20–26 Cursitor Street, London EC4A 1HY; tel 071 405 2088) for names of their members working in your area.

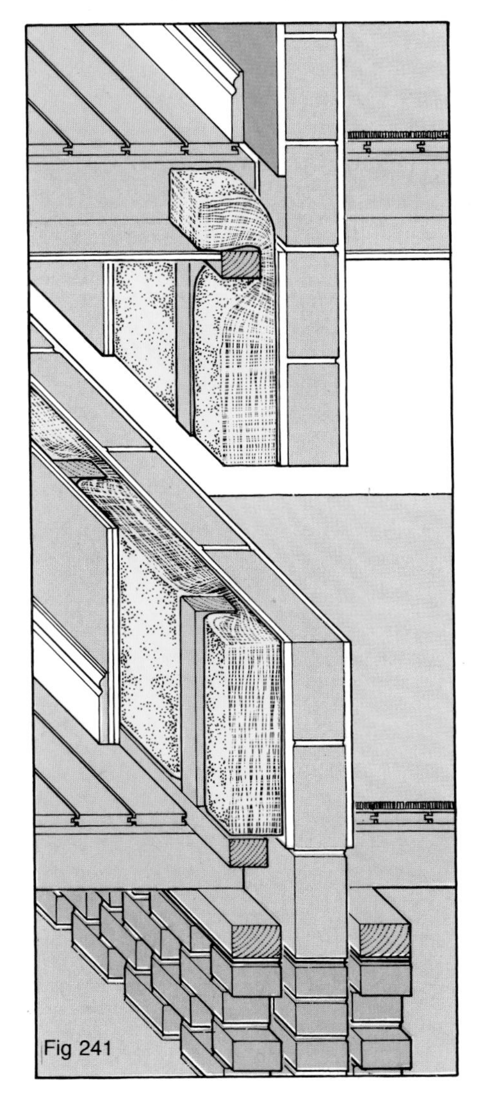

Fig 241 Build the framework as shown, with a head and sole plate linked by studs at 600mm (2ft) centres. Hang the lengths of insulation quilt behind the frame, and then clad it with plasterboard. Stagger joints between boards in the two layers.

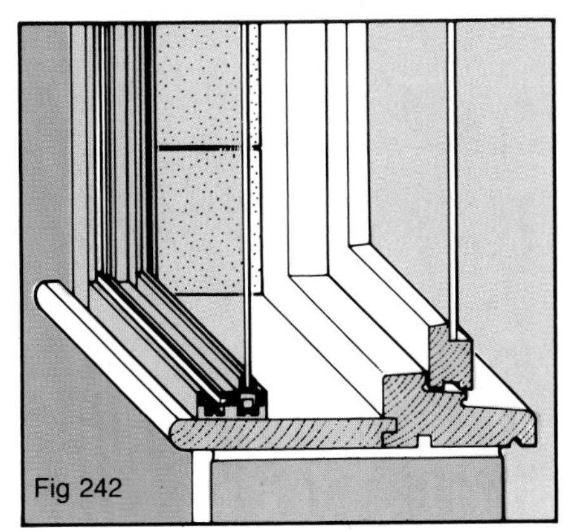

Fig 242 Site acoustic double glazing 150–200mm (6–8in) away from the existing window, and line the reveal with acoustic tiles. Keep both outer and inner panes closed to exclude noise, but open them if ventilation is required.

Fitting Ventilators

Good ventilation is as important for the health of your house as it is for that of its occupants. Without it, condensation can form on and within various parts of the house structure, and in time this can lead to serious problems. For example, it can encourage mould growth which will stain decorations and ruin stored clothes, and which can be very unhealthy to live with. As far as the building is concerned, the timber in suspended ground floors can be attacked by dry rot, which can be extremely expensive to eradicate. Brickwork can be damaged by frost, and damp masonry is a very poor insulator, which will encourage unnecessary heat losses. Insulation in lofts and walls can become saturated, causing damage to ceilings and wall linings and again encouraging heat loss.

The solution to the problem is to ensure that ventilators are fitted at certain key points around the house, and to give them a helping hand by extracting moist air mechanically from the main steam-producing troublespots, which are kitchens and bathrooms.

What to do

Start by checking the ventilation of underfloor voids to make sure that air-bricks are clear of obstructions. There should be air-bricks about every 2m (6ft 6in) on opposite sides of the void; install extra ones if they are more widely spaced (or are absent altogether).

Fit extractor fans to serve kitchens, bathrooms and windowless utility rooms or WCs. You can install the fan in a window pane, a hole in the house wall or even on the ceiling, in which case the extracted air must be taken to the outside through ducting run in the ceiling void or loft. Never discharge air from a fan directly into the loft space or into a disused chimney flue; serious condensation could result if you do.

Lastly, improve background ventilation in other rooms by fitting trickle ventilators to the heads of wooden window frames. Drill a line of 10mm (3/8in) diameter holes at 20mm (3/4in) centres through the frame, and then fit the inner and outer parts of the vent over them.

What you need:

- air-bricks
- cavity liners
- mortar
- bricklaying tools
- extractor fan(s)
- ducting and grille
- electrical tools
- trickle ventilator(s)
- carpentry tools

CHECK

- that rooms containing fuel-burning appliances have enough ventilation to guarantee their safe operation. Check with your fuel supplier if you are unsure about the ventilation requirements of your appliances

TIP

Wire extractor fans up to a humidity detector so they will operate automatically whenever moisture levels rise.

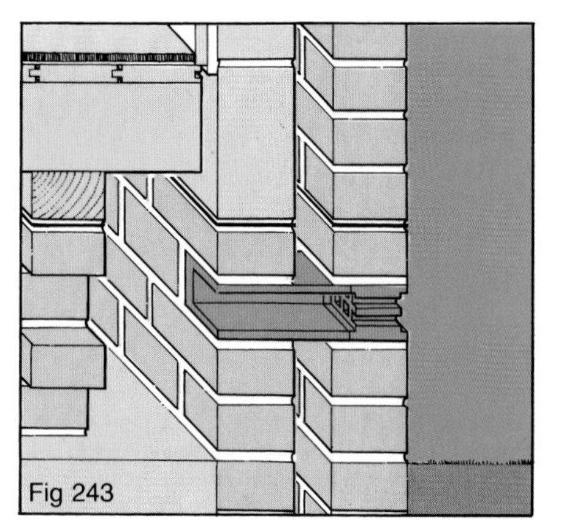

Fig 243

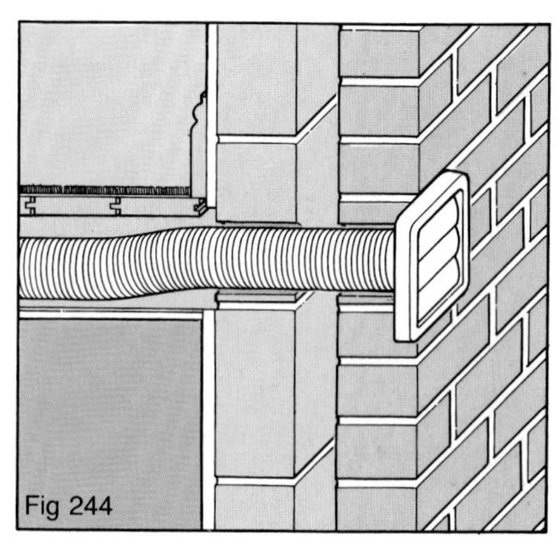

Fig 244

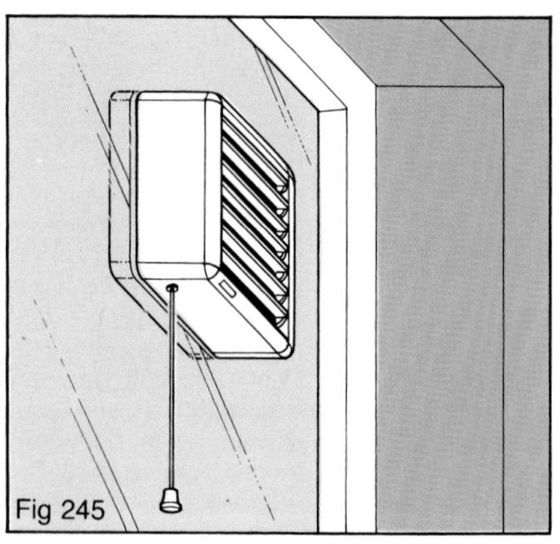

Fig 245

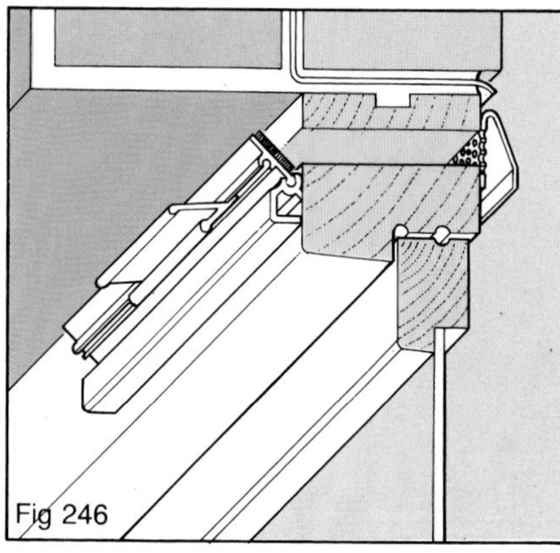

Fig 246

Fig 243 Fit extra air-bricks (with cavity sleeves in cavity walls) to ensure adequate ventilation of underfloor voids.

Fig 244 Use ducting to carry air to a discharge grille if the fan is remote from an outside wall. Never discharge air from a fan into the loft or a disused flue.

Fig 245 Fit window, wall or ceiling-mounted fans in kitchens and bathrooms to remove moist air before it can spread to other parts of the house and cause condensation.

Fig 246 Fit trickle ventilators to the heads of wooden windows to improve background ventilation in other rooms in the house.

EXPENSIVE JOBS

However competent a do-it-yourselfer you are, there will be some jobs that are either too big for you to tackle or require professional skills and equipment. Examples include such major tasks as retiling a roof (not necessarily beyond your skill, but likely to take too long if tackled on a DIY basis), laying hot asphalt, replacing a damp course or installing cavity-wall insulation. In all these cases, you will have no choice but to call in an outside contractor to do the work for you. This is a move that many people regard with some trepidation, since finding good, reliable tradesmen can be difficult and choosing the wrong firm could result not only in a bodged job but in considerable financial loss as well. Here are some general guide-lines to help you minimize the risk of picking a cowboy, and an explanation of what is involved in employing professionals for four individual projects.

Fig 247 (*above*) Re-roofing a house is a big job that most do-it-yourselfers would prefer to leave to a professional.

Finding a Contractor

Once you have decided you need to call in a professional, your first step is to get people with the skills you require to visit the site and give you a firm quotation for the job.

Personal recommendation is by far the best and safest way of finding someone suitable. If a firm has already carried out work for friends, relatives or neighbours you will be able to get a first-hand account of its performance and even be able to check up on the standard of workmanship.

If this does not work, your next step is to take a walk or a drive round your area, looking for signs of someone carrying out the sort of work you want done. Many firms now put up a sign outside the site they are working on (or park their vans close by), and will not mind if you approach them. You can do this directly, or, alternatively, you can telephone the number on the sign/van.

You can also approach householders directly if it is obvious that they have recently completed work such as roofing or damp-proofing. Most people are only too happy to show off a job well done and to put you in touch with the contractor who carried out the work.

Next, try your Yellow Pages or Thomson Local telephone directory. Both list local contractors by trades, and many of the display advertisements not only give more details of the sort of work undertaken, but may also reveal whether the firm is a member of a relevant trade association (*see* below). With this method, it is well worth asking the firm about other jobs it has done locally. Any company worth its salt will be pleased to put you in touch with satisfied customers.

Your last method of contact with local contractors is via the various professional and trade associations to which many reputable companies and individuals belong. These associations will give you the names and addresses of their members working in your area, and some offer other back-up services such as guarantees and arbitration schemes which may be worth knowing about. Membership of such a body is generally a good sign (many require evidence of several years' trading and satisfactory accounts before granting membership), but it is wise to check firms who claim membership with the body concerned as some firms simply 'borrow' logos and claim membership to enhance their image. For more details about individual trade associations that you could contact, *see* page 94.

Quotations

Getting Quotations

Once you have contacted someone who sounds interested in carrying out whatever work you want done, your next job is to explain clearly what the job involves and to find out as precisely as possible what it's going to cost you, when work can start and how long it will take to complete. For projects such as reroofing, damp-proofing and insulation it is essential not to rely on verbal agreements, but to ensure that everything is in writing. This can save a lot of arguments, and will also help a court to sort a dispute out if things go badly wrong. First, make sure you understand the meaning of the following terms, so you know what you are asking for and what the contractor intends you to get.

Estimates are just a guess as to the rough cost of the job. They are not legally binding.
Quotations are firm offers to carry out a specified job for an agreed price. A quotation for a simple job should include details of material costs.
Tenders are also offers to carry out specified work for a named price, but are understood to involve an element of competition with other contractors.

Most contractors will want to make a site visit to assess the scale of the job involved before even giving an estimate. Explain in as much detail as possible what you require, and tell him you require a firm quotation for the work, plus details of when he will start and how long the job will take.

Ask at this point whether he or the firm is registered for VAT, and if so whether VAT is payable on the work you are having done. Generally speaking, you do not have to pay VAT on work involving the construction, alteration or demolition of a building, but VAT is payable on repairs and maintenance. If he is not a registered VAT trader (with a registration number printed on his notepaper), he cannot charge you VAT on work he does for you.

Always get at least two quotations for the job, and more if you can. This allows you to compare terms as well as prices before making your choice. Impossibly high quotes rarely mean you'll get a top-quality job; they are the contractor's way of saying he does not want the job, but will do it if you are prepared to pay a silly price.

Assessing Quotations

Once you have received the quotations, study them carefully. The amount of detail given will vary, but the ideal quotation should cover the following points:
1 a description of the work to be carried out, preferably presented as a detailed list setting out all the stages involved (known as a schedule).
2 details of particular materials or fittings to be used, and who will supply them.
3 who will be responsible for obtaining any official permission needed for the work.
4 when the work will start.
5 when the work will be completed.
6 who will be responsible for insuring the work and materials on site (professional contractors have both employer's liability and public liability insurance).
7 whether subcontractors will be employed, and for which parts of the job.
8 how variations to methods, timing or cost will be agreed.
9 the total cost of the work.
10 when payment will be required.
These details form part of the contract between you and the firm you decide to employ, so it is important that they are discussed and dealt with now, to prevent arguments later. Some contractors may include them on a standard form of contract sent with the quotation, or may print their terms and conditions on the reverse of their quotation. In either case, read them carefully; now is the time to discuss any clauses you do not want to apply.

Once you have received quotations from the firms you approached and made your choice, write and accept the quotation and, as a courtesy, also notify unsuccessful applicants that they have not got the job.

You now have a contract between yourself and the contractor. In most circumstances, there is no reason to suppose that anything will go wrong but, if it does, tackle it immediately so things can be put right. Mention problems verbally first of all, and if this does not resolve matters, follow up with a letter outlining the nature of your complaint and requesting specific action to correct it. Always keep notes of any discussions you have, and copies of any letters you send, in case a dispute cannot be resolved and you have to go to independent arbitration.

CHECK
- that your contract covers the following points (where relevant):

1 A detailed description of the work to be carried out
2 The projected start and finish dates
3 What will happen if work overruns the projected date in terms of compensation
4 That the contractor will be responsible for clearing up the site
5 That the contractor will comply with the requirements of all relevant rules, regulations, laws and bye-laws
6 That any variation to the work will be agreed in writing by both parties
7 That the contractor will arrange third party and employer's liability insurance
8 That the contractor will put right any damage he causes
9 That the contractor will put right any defects in his work which occur within a set period after completion
10 How payment will be made for the work
11 That an amount will be retained until the end of the period agreed for the correction of defects

There are four main weatherproofing and insulation jobs you may prefer to leave to professional installers: tiling (or slating) a pitched roof, resurfacing a large flat roof, injecting a chemical damp course and getting your cavity walls insulated (which you cannot do yourself anyway). Here is a brief look at some of the points to bear in mind when planning to have these jobs carried out.

Tiling a Pitched Roof

Your roof may have deteriorated with age and neglect to the point where local repairs are no longer the solution, and will have to be stripped and replaced completely. The first thing to establish is whether the covering or the fixings are at fault. For example, many slate roofs suffer from what is known as nail sickness; the nails holding the slates in place corrode and rust through, so the slates slip out of position and allow water to penetrate. It may be possible to strip the roof and reuse all but a few of the original slates, a course of action which will obviously cost a lot less than having new slates fitted. Similarly, tiles can often be reused if failure is due to rot in the tiling battens rather than physical damage to the tiles themselves. Even if large-scale replacement has to be considered, it may be possible to reuse a proportion of the existing tiles or slates – to cover say the front face of the roof – with new materials being used at the rear of the property.

If complete replacement is called for, choose what you use with some care, on two grounds. The first is looks. Always replace like with like if possible, to maintain the original appearance and character of the building. A dramatic change can be quite ruinous, as proved by the innumerable thoughtlessly reroofed Victorian terraces visible up and down the country. Slates were the standard roofing here, and if new slates are too expensive then synthetic slates can be specified instead. The second point to remember is weight. It is vital not to overload the roof structure by using a tile that is heavier than the originals.

If you are having the roof stripped, take the opportunity to have the underfelt and battens replaced, especially if they are in poor condition. It is a waste of money to put new tiles over old unsound materials.

Fig 248 (*above*) Resurfacing a flat roof with hot asphalt is definitely a job for a professional.

CHECK
- that firms you employ for large-scale projects are members of an appropriate trade association (*see* page 94 for more details)

Fig 249 (*far left*) Full-scale roof tiling is too big a project for the average do-it-yourselfer to tackle single-handed.

Resurfacing a Flat Roof

If you have a large flat roof that needs replacing – on a home extension, for example – then it is worth considering alternatives to the traditional three-layer felt treatment. The first is asphalt, which is applied hot and trowelled out to form a seamless surface that is both more durable and more attractive than felt. The second is to use one of the new high-performance membranes or polymeric systems, which offer excellent performance and long life.

As with pitched roofs, be prepared to have to replace some or all of the decking if the failure of the roof covering has allowed moisture to penetrate the roof structure. This is also the ideal opportunity to cut heat losses through the roof, either by laying insulation on top of the roof decking before the new covering is applied if the decking is in good condition, or by using a composite decking material if it is not. These composite materials consist of foam insulation bonded to exterior-grade plywood, and include a vapour barrier as well.

Damp-Proofing Walls

If your house suffers from rising damp and the cause is not a simple damp bridge (*see* page 54), then the damp-proof course (DPC) has probably failed. The quickest and least expensive method of installing a new damp course is to inject special silicone liquid into the brickwork all round the house, along the line of the existing DPC. It is possible to hire the equipment and inject the liquid yourself, but there are several reasons why this is best left to a specialist contractor.

Firstly, the job can be technically quite difficult to carry out properly, especially if the house has cavity walls or the damp has also affected internal load-bearing walls. Since some of the brickwork you are trying to treat is concealed, you therefore have no way of knowing whether you have completely saturated it or formed a continuous damp barrier. A professional installer can ensure this.

Secondly, a professional firm will give you a guarantee for its work, although this is only worth having when something goes wrong if the firm is still in business. Therefore, employ a firm which is a member of the British Wood Preserving & Damp-proofing Association (*see* page 94).

Fig 250 (*above*) Installing a chemical damp-proof course can be carried out on a DIY basis, but whole-house treatment is best left to a professional contractor.

Cavity-Wall Insulation

If you plan to have your cavity walls insulated, you must first of all apply to your local authority for Building Regulations approval for the work. The building inspector will advise you on the suitability of the various materials used for your particular location (there are restrictions on the use of urea-formaldehyde foam in exposed locations, for example). He will also want to ensure that the firm which will be carrying out the work is approved by the British Standards Institute or is installing a product that is certificated by the British Board of Agrément (BBA).

All the materials used are injected or blown into the cavity through holes drilled in the outside leaf of the wall. Make sure that the firm checks whether your wall cavities are open or closed at eaves level; if they are open, insulation – especially foam – can find its way into the loft. Problems can also arise at other points where the cavity is bridged and not sealed. In particular, care must be taken to keep air-bricks and ventilators open.

If you live in a semi-detached house, try to persuade your neighbour to have his walls insulated at the same time. You will get a price reduction from the installer for treating two houses at the same time.

Fig 251 (*far left*) Cavity-wall insulation is pumped or blown into the cavity from the outside – definitely not a DIY job.

FACTS AND FIGURES

This section is intended as a handy reference guide to the range of roofing, weatherproofing and insulation products you will need in order to carry out the various jobs described earlier in the book. It will help you to see at a glance what is available, and in what sizes or quantities, so you can plan your requirements in detail and draw up shopping lists for individual jobs.

Lastly, on page 94, there is a detailed glossary of all the terms used in the book, plus a list of useful addresses.

Tiles and Slates

Roof tiles come in a wide range of shapes and sizes, but the most common are described as 'plain' tiles and are made from fired clay or concrete. Standard modern plain tiles measure 265 × 165mm (10½ × 6½in), and are between 10 and 15mm (3/8 to 5/8in) thick. Shorter tiles, for use in eaves courses, are a standard 190mm (7½in) long, while wider tile-and-a-half tiles for use at verges measure 265 × 250mm (10½ × 10in). When laid to a standard head lap of 75mm (3in), coverage is approximately 64 tiles per sq m (6 per sq ft). The weight of the covering is approximately 70kg per sq m (14lb per sq ft).

The standard half-round ridge-tiles usually used with plain tiles are also made in clay and concrete. Clay types come in 300 and 455mm (12 and 18in) lengths, in a range of widths from 150 to 250mm (6 to 10in). They are made in half-round and square angle profiles, the latter in a range of angles from 75° to 135°. Concrete half-round ridge tiles and third-round hip tiles are generally 455mm (18in) long and 250mm (10in) wide.

Interlocking concrete tiles are larger than plain tiles. Common sizes include:

- 430 × 380mm (17 × 15in), with a coverage of 8.2 tiles per sq m (7 per sq yd).
- 420 × 330mm (16½ × 13in), coverage 9.7 per sq m (8.1 per sq yd).
- 380 × 230mm (15 × 9in), coverage 16.5 per sq m (13.8 per sq yd).

The weight varies according to the tile profile, but is commonly around 45kg per sq m (9lb per sq ft), so replacing a plain-tiled roof with concrete interlocking tiles should pose no problems of overloading.

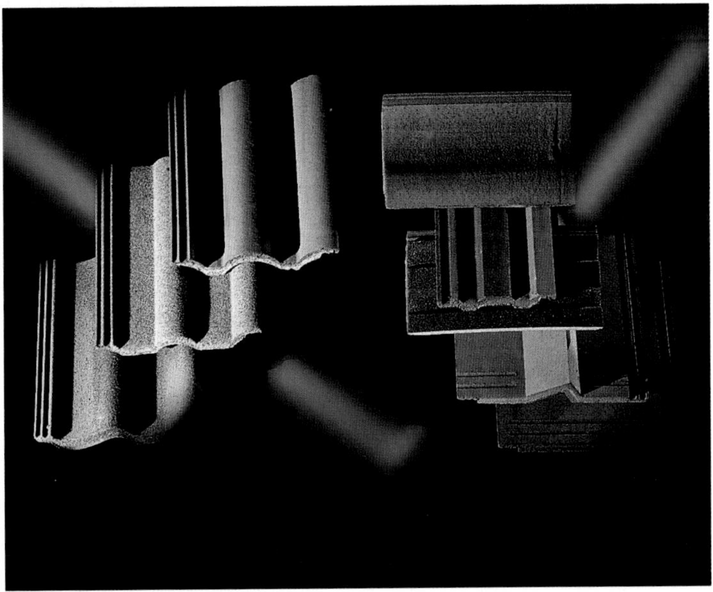

Fig 252 (*above*) Concrete tiles are available in a wide range of profiles and colours, and modern types are less prone to bleaching and fading than earlier versions.

Natural slates have over the centuries been cut in a wide range of different sizes, with an equally bewildering assortment of names. The largest was known as the empress and measured 660 × 405mm (26 × 16in), while the smallest was simply called a unit and measured 255 × 150mm (10 × 6in). In between came such delights as small duchesses, wide countesses and narrow ladies. Thankfully a range of eight or nine common standard sizes is now used, ranging from 610 × 305mm (24 × 12in) down to 355 × 180mm (14 × 7in). Thicknesses are standard (5mm/1/5in), heavy (6.3mm/¼in) and extra heavy (8.5mm/⅓in).

Coverage and weight per unit area obviously depends on the slate size. Coverages range from 40 per sq m for the smallest standard slates down to around 12 for the 610 × 305mm size. Weights range from 28kg per sq m (5.7lb per sq ft) for large slates up to 34kg per sq m (7lb per sq ft) for the smallest, which is comparatively light as far as roofing materials go. Therefore, you need to take care when replacing a slate roof with any type of tile to check that the roof structure is strong enough.

Imitation slates are usually available in a smaller range of standard sizes, with 600 × 300mm (24 × 12in) the commonest. Coverage at this size is about 12 per sq m; the weight ranges from 20 to 25kg per sq m.

Flat Roof Materials

Roofing felt is the basic ingredient for most do-it-yourself work on flat roofs. Felts are divided into five different classes, numbered from 1 to 5, and this classification is followed by a code letter that indicates the surface finish; for example, B denotes a plain sanded finish and E a mineral granule finish. All types should be marked as confirming to British Standard BS747:1977.

Felts are also graded by weight, with the figures denoting the weight of a roll 10m (33ft) long and 1m (3ft 3in) wide. The lightest type is 14kg felt, used for single-layer felting on outbuildings and the like. Heavier felts come in a range of weights, the commonest being 18, 25 and 38kg. Note that while 10m rolls are standard, longer rolls (usually 20m) are also available; these are still designated by the weight of a 10m long roll.

Class 1 felts have a base of animal and vegetable fibres saturated with bitumen, and are the cheapest type. They are used as single-layer coverings on outbuildings, as part of a built-up roofing system and as underlay felt for tiled and slated roofs. They are identified by a white line along one edge of the underside of the felt (except for the 14kg weight, which has no colour coding).

Class 2 felts have an asbestos fibre base, again saturated with bitumen, and are used to meet the fire resistance requirements of the Building Regulations. They are colour-coded with a green line.

Class 3 felts have a glass fibre base which does not have to be saturated with bitumen before the top coat is applied. They are stronger and more flexible than class 1 or 2 felts, so give better performance in built-up roofs. They are colour-coded red. Perforated types, designated class 3G with a grit underside and sand finish, are for use as the first felt layer on roofs when partial bonding to the decking is required.

Class 4 felts are used as underlays beneath asphalt and metal roof surfaces, and are known as hair or sheathing felts. They usually come in longer 25m (80ft) rolls and are 810mm (32in) wide. They are not colour-coded.

Class 5 felts have a base of polyester fibres, and are saturated with bitumen like class 1 and 2 felts. They are the strongest and most durable of the BS747 felts, and are the best type to use for major roofing projects such as flat-roofed extensions. They are colour-coded blue.

There are also several high-performance felts on the market which are not classified by the British Standard. However, many have been tested and approved by the British Board of Agrément (BBA), and will be included in the latest revision of the standard.

Sheet Roofing Materials

The commonest sheet roofing materials you are likely to be using are corrugated types – galvanized steel, 'asbestos cement' or some type of plastic. In fact, asbestos-cement types have now been superseded by types free from asbestos, for obvious safety reasons. All are mainly used on outbuildings and for roofing simple lean-to structures such as carports.

Sheets of all three types are generally available with standard 75mm (3in) corrugations in a range of sheet sizes. Plastic sheets, especially the cheaper PVC types, are also made with a smaller 32mm (1¼in) profile. These are laid with a side overlap of two corrugations, while the larger profiles need only one. Remember to allow for this when estimating the number of sheets needed to cover a given area.

Fig 253 (*below*) Roofing felts are classified according to the type of fibre used as their base material, and come in a range of weights for different jobs. All come in standard 10 × 1m rolls.

Damp-Proofing Materials

At first glance, the wide range of damp-proofing products on the market can appear very confusing, particularly since some products claim to be able to tackle several problems. They fall broadly into the following categories. The notes explain briefly what each product does and give details of coverage where appropriate.

Bituminous waterproofers are black liquids which are simply brushed on and left to dry, and form a waterproof coating that will cure pinhole leaks in surfaces such as flat or corrugated roofs. They are water-based products which dry fairly quickly and can be washed off tools and hands with water so long as they are still wet. Coverage: average 1.6sq m (17sq ft) per litre per coat.

Heavier-duty solvent-based waterproofers are also brush-on treatments that are black in colour, but are based on advanced resin formulations rather than bitumen and some may need special solvents for cleaning up brushes. They are ideal for waterproofing flat roofs in poor condition, and can be used with a reinforcing fabric on pitched slate roofs as well as flat or corrugated ones as a cure for overall failure of the roof covering. Coverage: 0.75sq m per (8sq ft) per litre per coat.

Rubberized bitumen emulsions are generally used to form damp-proof membranes (DPMs) in solid floors and basements, but they are also useful for jobs such as roof repair work, sealing gutters and so on. Coverage: 1–1.5sq m (11–16sq ft) per litre per coat.

Bituminous paints are not the same as bituminous waterproofers. They are intended for protecting exterior metalwork – gutters, downpipes and so on – and not as damp-proofers. The coverage depends on the roughness of the surface.

Black bitumen mastics are used for filling cracks and holes in flat and corrugated roofs, flashings, gutters and downpipes.

Cartridge mastics come in various types and are ideal for filling gaps between door and window frames and the surrounding masonry, sealing round roof-lights and ventilators, and for repairing gutter and downpipe leaks. They are generally white,

Fig 254 (*left*) Liquid damp-proofers are widely used for sealing leaking flat roofs. It is best to use throwaway brushes to apply them.

grey or cream in colour and fit in a standard mastic gun.

Silicone water repellents are clear liquids formulated for waterproofing exterior masonry – brickwork and rendering – that has become porous and is allowing damp to penetrate. You can also use them to provide a damp-proof course (DPC) in walls where the original DPC has failed, by injecting the liquid into the walls under pressure. Coverage: 2–4sq m (22–44sq ft) per litre per coat, depending on the wall porosity.

Repair tapes come in several disguises. Flashing tape is one of the most useful; as its name implies it is used to patch torn or porous flashings between roofs and walls, on chimney-stacks, in valley gutters and so on. Most have a metallic finish, and are bonded in place with a special primer. Waterproofing tapes are mainly used for repairing cracks and leaking glazing bars in glass roofs, and for lining leaky gutters, and there are clear types which are useful for making less visible repairs to glass and corrugated plastic.

The various insulation materials you are likely to be using can be placed in one of four categories: blankets, batts, loose-fill and board. In addition, you may need tank and pipe insulation in specific sizes. Here is how each type is sold.

Blanket insulation comes in rolls, in a number of different sizes; generally the thicker the blanket, the shorter the roll. Most blankets for do-it-yourself installation are sold in rolls between 370 and 400mm (around 15in) wide, to fit between joists at standard spacing. Common roll sizes are 8 and 5m (26 and 16ft) long for blankets 100mm (4in) thick, and 4 and 2.66m (13 and 8ft) long for blankets 150mm (6in) thick. For loft insulation, estimate your needs by measuring the eaves-to-eaves width of the loft and multiply that by the number of joists.

Insulation batts are sold in packs, usually containing 10 batts measuring around 1140 × 465mm (45 × 18in), so a pack will cover an area of about 5sq m (54sq ft).

Loose-fill insulation comes in bags. Mineral-fibre types are usually sold by

Fig 255 and 256 (*top and bottom left*) Cartridge mastics and paints can be used for a wide range of repair and weatherproofing jobs.

Fig 257 (*top right*) Blanket insulation is sold in rolls; the length decreases as the thickness rises. Loose-fill materials come in bags, sold by weight (mineral-fibre types) or volume (vermiculite).

Fig 258 (*bottom right*) Board insulation of all types comes in standard 2440 × 1220mm (8 × 4ft) sheets.

weight, with a 10kg (22lb) bag covering up to 2.5sq m (27sq ft) when laid to a depth of 100mm (4in). Vermiculite is sold by volume; one bag typically covers an area of around 1sq m (11sq ft) when laid at a depth of 100mm.

Board insulation, including plasterboard of various types and also rigid polystyrene slabs, is sold in standard-sized sheets, usually 2440 × 1220mm (8 × 4ft), although smaller sizes of polystyrene slab are widely available, especially the 1220 × 610mm (4 × 2ft) size.

Tank and Pipe Insulation

Tailor-made insulation jackets are widely available for the commonest sizes of hot cylinder (36 × 18in and 42 × 18in) and for cold-water storage and feed-and-expansion tanks (25 or 50 and 4 gallons respectively). Other sizes may have to be ordered.

Pipe insulation bandage comes in 3m (10ft) rolls, while pipe sleeving of all types is usually sold in packs of five or ten 1m or 2m (3ft 3in or 6ft 6in) lengths.

Draught-proofing materials are sold in four main forms, according to type.

Individual excluders for door bottoms and letter-box openings are sold singly. The former are generally 915mm (36in) long, and are designed to be cut down to match the width of the door or threshold to which the excluder will be fitted. If you have wider-than-usual doors, you may have to order a commercial-type excluder from a builders' merchant.

External door sets consist of three lengths of draught excluder, usually with a brush pile or flexible rubber seal set in a rigid plastic or metal moulding, which are cut to length as necessary and then fitted round the sides and top of the door frame.

Strip excluders of various types, including self-adhesive foam and V-seal excluders, are sold in rolls of various lengths – commonly 5, 10 and 25m (16, 33 and 80ft).

Mastic-type excluders are sold in cartridges which fit into standard mastic guns. Coverage depends on how thickly the mastic is applied.

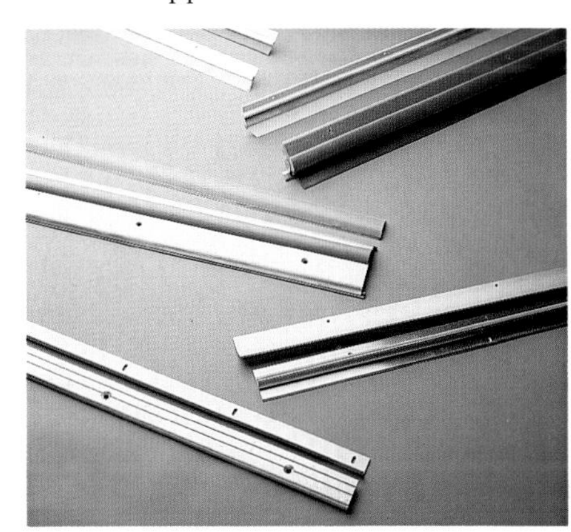

Fig 259 (*top left*) Ready-made jackets are available for common sizes of hot cylinder and cold tank.

Fig 260 (*bottom left*) Pipe insulation comes in rolls or in standard 1m or 2m lengths.

Fig 261 (*top right*) Excluders for door bottoms are usually 915mm (36in) wide, and have to be cut down to suit the door width.

Fig 262 (*bottom right*) Excluders for round door and window frames are sold as door sets or in roll form.

Glossary

Abutment The junction between a roof and a wall rising vertically above it.
Barge-board Length of timber or other material used to finish off the roof edge at gables, nailed to the ends of the roof timbers.
Battens Slim horizontal pieces of wood nailed to pitched roof rafters over the underfelt to act as a fixing point for tiles or slates.
Bonnet tile V-shaped tile used to cover the hip on plain-tiled roofs.
Built-up roof A flat roof surfaced with two or three layers of roofing felt.
Capsheet The final layer of roofing felt on a built-up roof.
Cold roof A flat roof constructed with insulation below the roof decking.
Damp-proof course (DPC) Waterproof barrier built into all walls rising off the ground to prevent damp from rising in the masonry. In older homes, slate was often used, but modern homes have DPCs of bituminized felt or plastic. Faulty DPCs can be cured by injecting special chemicals into the brickwork.
Double lap Tiles laid so that each one overlaps two layers of tile beneath it.
Downpipe Vertical pipe carrying rainwater from the gutters down to ground level, where it is discharged into a gully and runs on to a drain or soakaway.
Eaves The lowest part of a pitched roof or the overhanging edge of a flat one.
Fascia Vertical plank nailed to the cut ends of roof rafters at the eaves. Where the eaves overhang, a horizontal soffit fills the gap between fascia and wall.
Flashing Strip of waterproof material – usually metal or felt – used to waterproof the join between a roof and another adjacent surface (usually vertical).
Flaunching The sloping mortar fillet round the base of a chimney pot, sealing it to the stack.
Gable The triangular part of a house wall at the end of a ridged roof.
Gutter Metal or plastic trough fitted at the eaves to collect rainwater and divert it into a downpipe.
Hip The sloping external angle where two pitched roof slopes meet, weatherproofed with hip or ridge-tiles.
Interlocking tile Single-lap tile with the side overlaps grooved so each tile interlocks with its neighbour.
Joist A horizontal beam directly supporting a floor or ceiling.
Mineral wool Material made from mineral fibres, used as loose-fill, blanket and slab insulation.
Mastic Non-setting filler used to seal gaps between building components such as frames and masonry.
Mortar Mixture of sand and cement with added plasticizer, used for bricklaying, pointing and rendering.
Nib Projection on the back of a roof tile, used to locate it on its batten.
Pitch The angle of a sloping roof.
Plain tile Clay or concrete roof tile laid with a double lap.
Pointing The finish given to the mortar courses in bricklaying. Several different profiles are used.
Rafter Sloping timber extending from the eaves to the ridge or hip of a pitched roof.
Ridge The highest point of a pitched roof.
Roofing felt Flexible fibrous sheet used to cover a flat roof, and as an underlay beneath tiles and slates.
Single lap Tiles laid so that each tile overlaps just one tile beneath it.
Slate Natural roofing material, quarried in blocks and split into thin sheets of various sizes, fixed in place by nailing to supporting battens.
Soffit Horizontal board fitted between the fascia and the house wall where the eaves overhang.
Valley The sloping internal angle where two pitched roof slopes meet, weatherproofed with valley tiles or metal sheet.
Verge The edge of a pitched roof where it meets a gable.
Warm roof A flat roof constructed with insulation above the roof decking.

Trade Associations

Builders
Building Employers Confederation, 82 New Cavendish St, London W1M 8AD
Tel: 071 580 5588.

Federation of Master Builders, 33 John St, London WC1N 2BB.
Tel: 071 242 7583.

Damp treatment firms
British Wood Preserving & Damp Proofing Association, 6 The Office Village, 4 Romford Road, Stratford, London E15.
Tel: 081 519 2588.

Double Glazing Firms
Glass & Glazing Federation, 44–48 Borough High St., London SE1 1XP.
Tel: 071 403 7177.

Insulation contractors
The Cavity Foam Bureau, PO Box 79, Oldbury, Warley, West Midlands B69 4PW.
Tel: 021 544 4949.

National Association of Loft Insulating Contractors, National Cavity Insulation Association, PO Box 12, Haslemere, Surrey GU27 3AN.
Tel: 0428 54011.

Roofers
National Federation of Roofing Contractors, 15 Soho Square, London W1V 5FB.
Tel: 071 734 9164.

Ventilation contractors
Heating and Ventilating Contractors Association (HVCA), Esca House, 34 Palace Court, London W2 4JG.
Tel: 071 229 5543.

National Association of Plumbing, Heating and Mechanical Services Contractors, 6 Gate St, London WC2A 3HX.
Tel: 071 405 2678.

Index

A
access equipment, types, 19
access towers, 22–3, 36, 38
acoustic double glazing, 83
air-bricks, 38, 39, 59, 79, 84
asbestos-cement roofing, 33, 90
attic rooms, insulating, 60–1

B
barge-boards, 25
  repairing, 40–1
battens, roof, 26–7, 28
blanket insulation
  types, 13–14, 92–3
  using, 56–7, 61, 64, 73
brickwork
  repairing, 37, 52–3
  weatherproofing, 37, 52
brush-strip draught excluders, 74–5
butterfly roof, 25

C
cast-iron rainwater systems, 45
cavity-wall insulation
  types, 14
  faults in, 47
  installing, 65–6, 88–9
chimney pots, replacing, 36
chimney-stacks
  checking, 8
  repairing, 36–7
condensation, 55
  curing, 57, 84
contractors, employing, 85–8

D
damp
  meter, 53
  penetrating, 47, 51
  rising, 47, 54, 88
damp-proof course (DPC), 47, 54
damp-proofers
  types, 12, 91
  using, 30, 33, 35, 43, 46, 52
doors, draught-proofing, 16, 74–6
double glazing
  acoustic, 83
  installing, 80–1
  sealed-unit, 80
  types, 17, 79–80
double-lap tiling, 26
downpipes, repairing, 44
draught excluders
  door, 74–6
  foam, 16, 74, 77, 79
  letter-box, 76
  windows, 77–8
draught-proofing materials, types, 16, 93

E
eaves, 25
  repairing 40–1
  ventilators, 59
external insulation, walls, 64–5
extractor fans, 84

F
fascia boards, 25
  repairing 40–1, 87–8
feed-and-expansion tanks, insulating, 67
felt
  flashings, 34–5
  roofing, 11, 30, 90
filler foam, 51, 71
fireplaces, blocking disused, 39
flashings
  checking, 8
  repairing, 33, 34–5, 37
flashing tape, 11, 29, 31, 33–5, 37, 43, 45
flat roofs, 25
  insulating, 62–3
  patching, 29
  resurfacing, 30
flaunching, repairing, 36–7
floors
  insulating, 72–3
  sound-proofing, 83
flues, capping disused, 39
foam draught excluders, 16, 74, 77, 79
frames, sealing, 50

G
gable roof, 25
gambrel roof, 25
glass-fibre insulation, 13, 56–7, 61, 64
glass roofs, repairing, 32
glazing beads, replacing, 49
gutters, repairing, 42–3, 46

H
hipped roof, 25
hoppers
  clearing blockages, 45
  repairing, 44
hot cylinders, insulating, 15, 69
humidity detector, 84

I
insulation
  cavity-wall, 14, 47, 65–6, 88–9
  checking, 9
  cost-effectiveness, 55
  flat roofs, 62–3
  floors, 72–3
  hot cylinders, 15, 69
  lofts, 13–14, 56–8, 60–1
  materials, 13–15, 92–3
  pipework, 15, 70–1
  roof slopes, 60–1
  sound, 82–3
  storage tanks, 15, 67–9
  walls, 14, 64–6, 88–9
interlocking tiles, 10, 26
inverted roof, 25

L
ladders
  roof, 19, 24
  safety, 20–1
  types, 19

Index

lead flashing, repairing, 34–5
leaded lights, repairing, 50
letter-box draught excluder, 80
loft insulation
  laying, 56–8, 60–1
  types, 13–14, 92–3

M
mansard roof, 25
mastics
  types, 13, 91, 93
  using, 29, 42, 51
mineral wool insulation, 13, 56–7, 61, 64
mortar flashings, 34–5

P
party walls, sound-proofing, 82–3
pipe insulation
  types, 15, 93
  using, 70–1
pitched roof, 25, 26–8, 60–1, 87
plasterboard, insulating, 60–1, 64
pointing, repairing, 37, 53
polystyrene insulation, 15, 60, 63, 67, 72
putty, replacing, 48–9

R
radiators, insulating behind, 66
rainwater systems
  cast-iron, 45
  repairing, 42–4, 46
reinforcing membrane, 31
repair tapes
  types, 12, 93
  using, 32, 33
ridge-tiles
  replacing, 26–7, 87
  types, 10, 89
roof
  checking condition, 7–8
  corrugated, 33
  flat, 29–31
  glass, 32
  pitched, 25–8, 60–1, 87
roof insulation
  installing, 56–63
  types, 13–14, 92–3
roof slopes, insulating, 60–1
roof tiles
  replacing, 26–7, 87
  types, 10, 89
roofing, felt
  types, 11, 90
  using, 30–1
rot, preventing, 47, 60

S
safety
  access equipment, 18, 20–4
  fire escape routes, 80
  ventilation, 84
sash windows, draught-proofing, 78
silicone mastic, 78
silicone sealer, 37, 92
single-lap tiling, 26
slab insulation, 13, 56–8, 61, 64, 73, 92
slates
  replacing, 28, 89
  types, 11, 28, 89
sliding windows, draught-proofing, 78
soffits, 25
  repairing, 40–1
solar-reflective paint, 30
sound insulation, 82–3

T
tank insulation
  fitting, 57, 67–8
  types, 15, 93
threshold draught-excluders, 16, 75
tiles
  interlocking, 10, 26–7, 89
  plain, 10, 26–7, 89
  replacing, 26–7, 87
  ridge, 11, 27, 89
trickle ventilators, 84

U
underlay, insulating, 73

V
V-seal draught-excluders, 16, 74, 77
valleys, 25
valley gutters, 25, 46
vapour barriers, 57, 60, 63, 64
ventilation
  air-bricks, 38, 39, 59, 79, 84
  lofts, 56–7, 59
  roof timbers, 60, 63
  safety, 84
  underfloor, 47, 54
ventilators
  eaves, 59, 63
  trickle, 84

W
wall insulation, types, 14, 64, 92–3
walls, exterior
  damp-proofing, 47, 51, 54, 88
  insulating, 64–6, 88–9
  sealing, 50
warm roof construction, 62
water tanks, insulating, 67–9
waterproofers
  types, 12, 91
  using, 30, 33, 35, 43, 46, 52
windows
  draught-proofing, 77–9
  sound-proofing, 83
  weatherproofing, 48–51